TRANSPORTATION PLANNING ON TRIAL

TRANSPORTATION PLANNING ON TRIAL

The Clean Air Act and Travel Forecasting

MARK GARRETT
MARTIN WACHS

SAGE Publications
International Educational and Professional Publisher
Thousand Oaks   London   New Delhi

For information address:

SAGE Publications, Inc.
2455 Teller Road
Thousand Oaks, California 91320
E-mail: order@sagepub.com

SAGE Publications Ltd.
6 Bonhill Street
London EC2A 4PU
United Kingdom

SAGE Publications India Pvt. Ltd.
M-32 Market
Greater Kailash I
New Delhi 110 048 India

Printed in the United States of America

Library of Congress Cataloging-in-Publication Data

Garrett, Mark.
    Transportation planning on trial: The Clean Air Act and
forecasting / Mark Garrett, Martin Wachs.
        pm.   cm.
    Includes bibliographical references.
    ISBN 0-8039-7352-7 (cloth: acid-free paper). — ISBN
0-8039-7353-5 (pbk.: acid-free paper).
    1. Transportation and state—United States. 2. Transportation—
United States—Planning. 3. Transportation, Automotive—Law and
legislation—United States. 4. Air—Pollution—Law and legislation—
United States. 5. Transportation, Automotive—Environmental
aspects—United States. 6. Traffic estimation—United States.
I. Wachs, Martin. II. Title.
HE206.2.G37   1996
388′.041—dc20   95-50228

This book is printed on acid-free paper.

96 97 98 99 00 10 9 8 7 6 5 4 3 2 1

Sage Production Editor: Michèle Lingre
Sage Typesetter: Christina M. Hill

Contents

ACKNOWLEDGMENTS   vii

1. INTRODUCTION   1

   Evolution of Travel Demand Models   5

   The Clean Air Act   9

   ISTEA   22

   Air Quality/Transportation Linkages   23

2. THE BAY AREA LAWSUIT   39

   The Bay Area Air Quality Plan   41

   Court Ruling on Liability Issues   53

   Summary   69

3. THE METROPOLITAN TRANSPORTATION COMMISSION'S
   NEW CONFORMITY ASSESSMENT PROCEDURES   81

The Transportation-Land Use Connection 83

MTC's MTCFCAST Computer Model 86

Deficiencies in the Modeling Process 94

Injunction Against Highway Projects 109

Revised Conformity Assessment Procedures 117

Court Ruling on Revised Conformity Assessment 129

Conclusion 137

4. IMPLEMENTING THE TRANSPORTATION
CONTINGENCY PLAN 149

Delaying Highway Projects 150

Remedying RFP Shortfalls 153

The "2131" TCMs 155

The Adequacy of the Contingency Measures 156

The Effect of the 1990 Clean Air Act Amendments 174

Court Ruling on Contingency Plan 179

MTC's Regional CO Compliance Response 183

5. ANALYSIS AND CONCLUSIONS 195

The Meaning of the Court Ruling for Transportation Planning
and Modeling 197

An Agenda for Transportation and Air Quality Analysis 219

Conclusion 222

APPENDIX A: CONFORMITY ASSESSMENT 229

APPENDIX B: CONTINGENCY PLAN 230

ABOUT THE AUTHORS 231

Acknowledgments

This book was prepared with the support of a grant from the University of California Transportation Center, for which we are extremely grateful. The findings and opinions expressed in this document are those of the authors and are not endorsed by the University of California or any organizations within the university. We express sincere thanks to Ms. Karen Kramer, law clerk on the staff of Judge Thelton E. Henderson, who read a draft of this book and offered helpful comments. We also thank Professor Allen Scott, former director of the Lewis Center for Regional Policy Studies at UCLA for his support in bringing this project to fruition, and Vanessa Dingley and Gloria Contreras for their technical and administrative support throughout the project.

MARK GARRETT
MARTIN WACHS

1

Introduction

Two recent laws are having major impacts on transportation planning and travel forecasting. Both the Clean Air Act amendments[1] and the 1991 Intermodal Surface Transportation Efficiency Act (ISTEA)[2] emphasize the role transportation systems play in attaining federally mandated air quality standards. These two pieces of federal legislation are among the most important landmarks in a decade-long shift of emphasis in regional transportation planning. The acts require new, detailed, accurate analyses of the potential impacts of transportation improvements on congestion, travel, and land use.[3] What is more important, the strategies developed to help meet regional air quality standards will affect public investments in the transportation network and public transit, which in turn will affect future regional development.

The Clean Air Act was enacted to protect and enhance the quality of the nation's air resources, to prevent and control air pollution, and to provide technical and financial assistance to state and local governments to develop pollution control programs. Congress recognized that the growth and complexity of air pollution due to urbanization, industrial development, and the increasing use of motor vehicles posed a danger to public health and welfare.[4]

1

The act sets national air quality standards and requires states to curb pollution from stationary and transportation-related sources. The 1990 amendments significantly expand the transportation planning requirements previously contained in the 1977 Clean Air Act. ISTEA is intended to develop
an economically efficient and environmentally sound national intermodal
transportation system. It gives state and local governments greater flexibility
in addressing transportation issues and provides funding for transportation
programs that contribute to meeting air quality standards.[5]

One concern that the new acts raise is that litigation over purported failures
to adequately assess the air quality effects of particular highway projects will
delay or cancel those projects and possibly others. Most travel forecasting
procedures currently in use cannot meet the requirements of these acts. Many
of the existing models were originally developed in the 1960s and 1970s and
have not been revised for more than a decade. Adapting them to meet today's
requirements will be necessary until newer models are operational.

Recent litigation in the San Francisco Bay Area pointed up many deficiencies in the current state of travel modeling and analysis, particularly in
assessing air quality impacts from transportation projects. Environmental
groups challenged the adequacy of transportation planning in the Bay Area
even though the Metropolitan Transportation Commission (MTC), the regional transportation planning agency, was at the time a national leader in
transportation planning methods. The lawsuit nevertheless contended that
planners had failed to consider the potential land use implications of new
freeway construction and funding decisions, failed to implement transportation control measures included in the regional transportation plan designed
to reduce traffic and achieve cleaner air, and failed to meet other federal
requirements of the 1977 Clean Air Act linked to the urban transportation
planning process.

Besides challenging the planning process as inadequate, the plaintiffs also
contended its content was flawed—that the standard models being used were
inappropriate to the task assigned them—and therefore the results could not
be relied on for accurate assessments of the air pollution impacts from
freeway construction. The federal court ruled in favor of the plaintiffs based
on the defendants' own planning commitments, and after a couple of years
of legal wrangling over how the defendants would comply with the court's
order, the parties reached a settlement in which the regional transportation
planners agreed to modify their procedures to better address air quality
concerns.

The Bay Area case, which is the subject of this book, is frequently cited as a threat to the current standard approach to regional transportation planning, because it holds planners to a heightened standard in meeting national air pollution goals. Although the case arose before the new legislation took effect, the court did address the impact of the 1990 Clean Air Act Amendments, particularly on existing pollution control strategies. Even prior to the amendments, the 1977 Clean Air Act required regional planners to incorporate air quality goals into their transportation planning process to receive federal highway project funding. The court ruled that the 1990 amendments reinforced those requirements and did not permit the agencies to relax their pollution control efforts. In brief, the decision held that existing planning methods failed to satisfy the 1977 act and by extension would also not be adequate to comply with the 1990 amendments.

The decision has major implications for future planning practice under the 1990 amendments and ISTEA, because the court held the defendant transportation planning agencies responsible for greater proficiency in predicting regional impacts from freeway construction and upgrading than most current modeling practice allows. This higher level of analysis has become mandatory for all states under the new legislation and the EPA regulations designed to implement it. The lessons drawn from the Bay Area litigation should serve as a guide to other jurisdictions that do not presently comply with the federal clean air standards. Unless better techniques can be devised to assess potential air quality impacts of new freeways and added traffic, we may expect increasing challenges in court to slow or halt new highway and mass transit projects.

The case also raised a number of important questions concerning the relationship between transportation system investment and regional growth. Plaintiffs argued that when new highways are planned to respond to forecasts of future land development, their eventual construction will itself change those land use patterns and that this feedback loop must be incorporated into the modeling process. Although the court ruled that it has never been shown that highway construction causes growth in a region that would not otherwise have occurred, it did hold that highway construction may indeed affect the distribution of economic development and residential population within a region. Regional planners in the Bay Area were required to consider these form-inducing effects of highway construction, a practice that may well be required elsewhere under the new acts.

In short, the case pointed out that regional transportation planning methods in widespread use around the country may be inadequate to fulfill several requirements of the 1990 Clean Air Act Amendments and ISTEA. As a response to the new demands being placed on travel forecasting by these acts, since 1992 the Department of Transportation (DOT), in cooperation with the Environmental Protection Agency (EPA) and the Department of Energy, has undertaken a major new Travel Model Improvement Program, in which it intends to spend $25 million over a 5-year period to devise a new set of transportation models. The principal goal of this program is to develop travel forecasting procedures that will meet the new legislative requirements. These procedures must include forecasting models that accurately assess the congestion and air quality effects of highway and transit improvements. Four areas of program activity currently underway are as follows:

- Near-term training and technical assistance for metropolitan planning organizations and state transportation departments
- Improvement of current models for use until better models are available
- Development of fundamentally new models that meet today's needs
- Major efforts to collect current, specialized data for model development.

This book explores some key issues raised in the Bay Area litigation and their implications for future transportation planning throughout the country. One of the authors, Professor Martin Wachs of UCLA, was appointed to be a neutral technical expert for the court during the course of the litigation and is also the chair of the review panel for the DOT's Travel Model Improvement Program.

The court's decision is already having a major influence on transportation planning methods. It also raises a number of issues for the future, such as the following:

- Will the case lead to more vigorous application of transportation control measures?
- Will the ruling result merely in improved technical models to be applied within the current planning process, or
- Will the new modeling approaches actually lead to new understandings of the land use/transportation nexus?
- If so, will the plans developed using these new planning methods actually move metropolitan areas toward new urban forms?

The answers to these questions will depend on how planners respond to the issues raised by the Bay Area litigation. Before the full discussion of the case beginning in Chapter 2, the next sections provide some background on the history of travel demand forecasting, on the federal Clean Air Act and its relationship to transportation planning, and on the newly mandated coordination between the act and ISTEA.

Evolution of Travel Demand Models

Until very recently, state and federal transportation investments have been primarily based on a policy of relieving traffic congestion, not reducing traffic levels or improving air quality. Models used to estimate vehicle usage on various roadway segments were designed primarily to determine whether additional lanes or segments would be needed to carry the expected traffic. As a result, they had to be adapted to meet the needs of pollution abatement, a task for which they had not been designed. At present, the state of modeling practice is simply not capable of producing the refined estimates of vehicular traffic and emissions needed to evaluate the effects of different control measures or alternative freeway networks.

The first travel demand forecasting models were developed in the late 1960s and early 1970s in support of large scale regional transportation planning studies then under way. There had been little construction of new roads and new housing during the Depression years and throughout World War II. After the war, a national effort emerged to accommodate anticipated growth in automobile travel that would result from the satisfaction of the expected boom in the demand for new automobiles. A flood of suburban housing construction benefited from housing assistance through the Federal Housing Administration and the Veterans Administration and provided further growth in the demand for autos and travel. Some Americans favored suburban housing construction, expansion of automobile manufacturing and acquisition, and the provision of the supporting road network not only to ensure jobs and homes for returning veterans but also as an economic stimulus that might prevent the postwar economy from slipping back into depression. Others saw the spread of the suburban single-family auto-oriented lifestyle as the extension of pre-Depression market trends. Some decried the destruction of vacant land and the creation of homogeneous suburban communities, but transportation officials concentrated on building the road network to make the suburban expansion possible. The federal transportation program, exemplified by the 1956 Highway Act, made available to the states as much as 90% of the money needed to construct new access-controlled multilane freeways or expressways, provided that the networks of roads were built in accordance with comprehensive regional facilities plans. Regional transportation study organizations were created in the 1960s and early 1970s to prepare these master plans of freeways and, while carrying out their work, the staffs of these transportation study organizations began experimenting with the early generation of computers that had come into use toward the end of the war and was then advancing rapidly.[6]

The transportation study organizations were considering policy questions that differ significantly from those their successor organizations must consider today, and it is important to understand that the nature of transportation forecasting models was shaped by these earlier planning requirements. In most instances, planners in the 1960s and 1970s were concerned with the configuration of a regional network of high-capacity freeways: What should be the size, spacing, and general pattern of the network for a particular region? Should, for example, the region adopt a rectangular grid of freeways intersecting one another at approximately right angles, or should the region adopt a set of radial highways concentrating on the downtown area like the spokes of a wheel near its hub? And if there was to be a radial network, should there also be one or more circumferential beltways to accommodate traffic bound for areas outside the regional core? The early transportation computer models were used to test broad alternatives such as these:

> Hypothetical highway networks were superimposed on existing and forecast development patterns, and the models forecast how many trips would originate and end in each geographic area and what the levels of congestion would be on each alternative road arrangement when those trips were accommodated.

> The costs of constructing more and more lane-miles of highways were weighed against the costs to the travelers of travel time, their vehicle operating costs, accidents, and so forth.

At the time these analytical methods were being perfected and disseminated—in part by federal initiative—little was known about the air quality consequences of transportation investments, so air quality simply was not addressed by the models at all. Indeed, the models were not even terribly sensitive to the possibility of providing public transit as an alternative to the automobile, because the emphasis of transportation planning efforts at the time was on highway construction. Highway location, spacing, and sizing were the overriding objectives of the planning process, and the early forecasting models were intended to be tools by which to accomplish this job.[7]

Because of the limited capacity of computers, and to a lesser extent the limited availability of data in the early days of regional transportation planning, the early transportation studies developed what planners today call a *sequential-independent* set of models, meaning that several models were used in a certain order but each model in the sequence was actually separate from the others. The forecasting task was divided into a series of stages, as illustrated in Figure 1.1, and the separate models were used to accomplish the tasks associated with each stage. The results of the first stage became the inputs to the second, the results of the second became the inputs to the third, and so forth. Beginning at the top of the figure, the *trip generation* stage of modeling took existing land use and socioeconomic data and estimated how many trips would originate in each geographic area or zone. *Trip distribution* was a separate modeling process that took as given the trips originating at each location or zone and divided them among the many zones in the region that could be the destinations for the travel. A separate *mode choice* procedure determined how many of the trips from each origin area to each destination area would be taken by automobile and how many by transit. Still another independent *traffic assignment* procedure determined how the automobile trips between a given origin and destination would be distributed among alternative roads available for those trips. Traffic assignment results gave rise to indications of which roads would be congested and to consequent testing of different road arrangements to see whether the forecast traffic could be accommodated more cheaply or with less congestion under a different arrangement of facilities.

Over the 20- to 30-year life of travel demand models, many new transportation policy issues have emerged, and transportation planners have adjusted the models so that they may be applied to the analysis of these new questions. One of these issues, central to the Bay Area lawsuit, is air quality. To estimate the extent to which future travel patterns will create concentrations of air pollutants at different locations, the already long sequence of independent models is actually extended even further. First, by adding an emissions model to the sequence and assuming a certain mix of vehicles in the regional

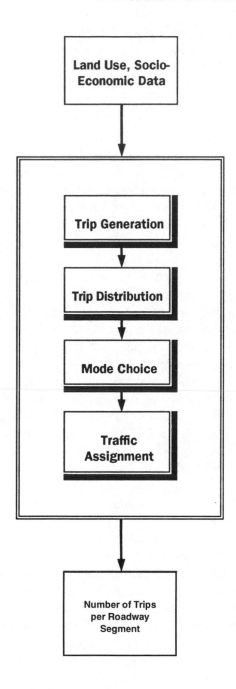

Figure 1.1. Standard Travel Demand Model

fleet, traffic flows are estimated to give rise to a particular set of pollutant emissions. Then, using a separate set of computerized models at least as complex as the travel demand models, emissions from traffic are added to emissions from stationary sources of pollution, such as power plants and factories, and the effects of climate, weather, topography, and sunlight are included to estimate the concentrations of pollutants at different locations and at different times of day.[8]

This standard procedure has been in use up to now to predict air quality impacts from transportation and other sources of air pollution to comply with the requirements of the federal Clean Air Act as discussed in the following section. The 1990 amendments and ISTEA, however, demand a more refined level of analysis of the expected emissions of criteria pollutants from vehicles than these models are capable of providing, a point made clear in the Bay Area litigation. More sophisticated techniques of travel forecasting will have to be developed to comply with these laws. Besides exposing the technical deficiencies in modeling practice, this case illustrated many of the problems that had arisen as a result of the failure of states to adequately integrate transportation and air quality planning under the Clean Air Act. It also reflected an ongoing debate between environmentalists and transportation planners over how best to address air quality issues. So that the readers can better understand the overall context in which the case arose, the next section briefly describes the history of the Clean Air Act and its relationship to travel forecasting.

The Clean Air Act

Congress passed the Clean Air Act in 1955.[9] It was mainly designed to foster cooperation between federal, state, and local government agencies in air pollution prevention and abatement efforts. Within less than two decades, however, national concern over the deteriorating air quality in many cities led to a substantial amendment called the Air Quality Act of 1967,[10] followed 3 years later by a major revision known as the Clean Air Amendments of 1970.[11] Automobiles are a major source of air pollution and it was recognized that constructing new freeways, although initially reducing traffic congestion, also tended to generate additional demand for travel. Among other things, the 1970 amendments gave states broad authority to adopt transportation control measures, commonly known as TCMs, to restrict automobile use and provide incentives for alternatives to single-occupant vehicle (SOV) trips, such as public mass transit. The amendment's congressional sponsors anticipated, perhaps naively, that the legislation would dramatically change the public's driving habits. Curbing traffic growth would prove more difficult than expected, however, and the emphasis would soon shift toward reducing automobile emissions at the source.

Air Quality Standards

A key part of the overall pollution control strategy was establishing new air quality standards. The 1970 amendments directed the EPA to set limits for atmospheric concentrations of certain key pollutants that have an adverse effect on public health or welfare. The limits, known as the National Ambient Air Quality Standards or NAAQS, establish minimum standards for these *criteria* pollutants, including carbon monoxide (CO) and ozone (O_3) emissions.[12] Exposure to ozone can damage the lungs and contribute to asthma, bronchitis, and emphysema. CO deprives the body of oxygen, which can impair the cardiovascular and nervous systems. These two pollutants are key components of photochemical smog.

Ozone is the main constituent of smog. It is actually a *secondary pollutant* formed when hydrocarbons (HC) and oxides of nitrogen (NOx), known as *primary pollutants* or *precursors,* react in the presence of sunlight.[13] These compounds are emitted from motor vehicles and industrial, commercial, institutional, agricultural, and domestic activities. The chemical reaction depends, among other things, on sunlight intensity, atmospheric conditions, and the relative concentrations of precursors.[14] Ozone is generally a regional problem, occurring predominantly during hot summer days. Because of the nature of the reaction, efforts to reduce both HC and NOx simultaneously have little effect on ozone formation. Therefore, strategies to control ozone primarily focus on reducing HC emissions.

The major source of CO is automobile exhaust. Although it can occur throughout a region, CO pollution is often localized. High concentrations may occur near areas of severe traffic congestion, known as *hot spots.* Unlike the difficulty with ozone, the problem is worse during cold weather.

The federal primary standards for ozone and CO are based on the levels of air quality necessary, with an adequate margin of safety, to protect the public health from these pollutants.[15] The standard for ozone is achieved when the expected number of days per calendar year with a maximum hourly average concentration above 12 parts per hundred million (pphm) is equal to or less than one.[16] The primary standards for CO are set for both 1-hour and 8-hour periods. Pollutant concentrations must not exceed 9 parts per million (ppm) on average over any 8-hour period, more than once per year. The 1-hour standard is 35 ppm.[17] In other words, pollution levels on the second worst air quality day of the year must be less than or equal to these ozone and CO standards to comply with the act. States were expected to develop plans to meet these standards.

State Implementation Plans

Under the 1970 amendments, each state was required to prepare and submit to the EPA for approval an air quality management plan known as the State Implementation Plan (SIP) for each designated *air quality control region* within the state.[18] The SIP had to include "emission limitations, schedules of compliance and such other measures as may be required" to attain and maintain the standards, including land use and transportation controls.[19] States had to show that the primary standards would be attained within 3 to 5 years.[20] In addition, the plan had to provide, to the extent necessary and practical, for periodic motor vehicle inspections and emission testing[21] and include provisions for revising the plan should the EPA later determine that it was "substantially inadequate" to achieve the federal standards.[22] If a state failed to produce an adequate plan, the EPA was given authority to prepare a Federal Implementation Plan (FIP) in its place.[23]

The 1970 amendments launched a major effort directed at lowering automobile exhaust. The legislation relied on two approaches for controlling transportation-related pollution: (a) reducing tailpipe emissions and (b) making changes to transportation systems. But because the federal government, not the states, regulates automobile emissions and because the EPA granted delays to automobile manufacturers to meet tailpipe standards, this meant that the states would have to rely more heavily on politically unpopular transportation controls to achieve the standards. Not surprisingly, few state plans contained such measures, and many other states simply delayed preparing their SIPs at all. Initially, the EPA granted the states additional time to submit adequate plans and approved a number of requested extensions for attaining the standards based on the unavailability of TCMs, but the courts rejected that practice, forcing the EPA to prepare FIPs for some areas and promulgate additional regulations for others.[24]

The EPA employed two main strategies to reduce automobile pollution. The first strategy consisted of promoting vehicle inspection and maintenance (I/M) programs and retrofitting older vehicles with pollution control devices. Inspection programs reduce tailpipe emissions by ensuring that all required smog systems on the vehicles are in place and operating properly. This strategy aims to reduce the amount of pollution coming from each vehicle. The EPA recognized, though, that for many areas, controlling automobile emissions alone would not result in attaining the NAAQS due to the increased traffic caused by population growth, suburban development, and higher vehicle use. Therefore, a substantial share of pollution reduction was placed on measures of the second type, mandatory transportation controls. TCMs are designed primarily to reduce the number of vehicles on the road, or the length or frequency of automobile trips being made. This strategy aimed to control the number of vehicle miles traveled (VMT) through restricting on-street parking, curtailing heavy-duty commercial vehicle use, and instituting mandatory parking fees, preferential bus/car pool lanes, computer car pool matching, bike paths, mass transit projects, and even gasoline rationing.[25] These measures were designed to close the gap left by stationary source controls and the federal motor vehicle control program. In addition to these direct transportation controls, the EPA also required implementation plans to include permit requirements for indirect sources of pollution, such as shopping centers, sports facilities, major roads, and airports, which attracted heavy automobile traffic.[26]

Many city and state officials became very concerned over these federal intrusions into what have traditionally been state and local decisions and a number of states filed lawsuits challenging the EPA plans and regulations.[27] In the end the EPA's actions proved too controversial and, in 1974, Congress began to curtail them.[28] The 1974 amendments specifically prohibited the EPA from requiring states to impose parking surcharges[29] or adopt bus/car pool lanes without a public hearing.[30] This policy shift away from placing heavy reliance on transportation control measures to improve air quality continued in the Clean Air Act Amendments of 1977, described next, although they have returned with force in the recent 1990 amendments. In place of these federally required land use and transportation controls, states were encouraged to adopt their own individual approaches for dealing with their air pollution problems.

The 1977 Amendments

Quite a few states failed to attain the air quality standards by the statutory deadlines despite the EPA's efforts. Therefore, in 1977, Congress again amended the Clean Air Act to encourage greater compliance with federal regulations while still allowing for additional growth to take place in already polluted areas.[31] The deadline for achieving the federal air quality standards was extended, allowing more time for federal automotive emissions standards to take effect. Convinced that mandatory transportation control measures were not workable, it further restricted their use. States could avoid existing SIP provisions calling for gas rationing or parking regulations,[32] or measures requiring bridge tolls or charges,[33] as Congress began shifting the emphasis on pollution control back toward improving technology and greater use of I/M programs, strategies that would not have such drastic impacts on driving habits. In addition, it required all SIPs to include preconstruction review of direct sources of air pollution,[34] but barred the EPA from requiring indirect source review.[35] Finally, the amendments initiated a new planning process for the states by requiring any state intending to construct new facilities that could cause NAAQS to be exceeded in any area to submit a revised nonattainment plan.[36]

Nonattainment Plan Requirements

The Clean Air Act Amendments of 1977 required the EPA to compile a list of those air quality control regions that did not then meet the national standards.[37] Part D of the Act, 42 U.S.C.A. § 7501 et seq., spelled out the plan requirements for these so-called nonattainment areas.[38] States had until January 1, 1979, to submit revised SIPs containing strategies to achieve compliance with the primary NAAQS.[39] The standards had to be met "as expeditiously as practicable" but no later than December 31, 1982.[40] Each nonattainment plan had to contain a comprehensive, accurate, current inventory of actual emissions from all sources[41] and was to expressly quantify the emissions *growth allowance* for each pollutant from the construction and operation of all major new or modified stationary sources of pollution.[42] The 1977 amendments did not explicitly require an analysis of projected growth in vehicle emissions or the effect of new highways on travel demand, though states were free to include more specific provisions in their SIPs. The plans had to contain whatever measures were necessary to meet the federal standards by the new deadline.[43] In addition, states had to maintain reasonable further progress (RFP) toward attainment in the interim period.[44]

To minimize the need for additional controls on vehicles and other existing sources of pollution, states also had to adopt a permit program covering all new major stationary sources.[45] The state could approve new sources only if emissions from all new sources did not exceed the growth allowance in the SIP or there were sufficient offsets provided so that emissions from all existing and proposed sources reflected RFP.[46] This program and the other SIP provisions had to impose sufficient controls to demonstrate that the state could in fact achieve the NAAQS as required by the act, given any anticipated increases in ambient pollution, for the SIP to meet the Part D requirements. Congress authorized the EPA to impose sanctions, including a construction ban on major new air pollution sources[47] and the withholding of federal funds for new highway projects,[48] on states that failed to submit adequate plans. Political considerations, however, often made it all but impossible for the EPA to do so.

To demonstrate attainment, a state had to show that by the legislative deadline pollution projected from all sources would be less than the level corresponding to the maximum allowable concentrations for each of the criteria pollutants. Each state prepared the required inventory of current emissions from all sources for the base year of the plan and estimates for future years, including the expected year of attainment. The states then used a mathematical formula to calculate how much projected emissions in the attainment year must be reduced to comply with the NAAQS. Any difference between the expected and target levels would have to be eliminated through control programs covering stationary sources or transportation sources or both. For transportation planners, this meant estimating not only the emissions generated by all vehicles expected on the road in the target year but also the expected reductions from all proposed control measures, including state vehicle inspection programs and TCMs. The modeling procedures described earlier were used to prepare emissions estimates, both with and without the proposed controls, to satisfy the state's attainment demonstration.

States were expected to incorporate all "reasonably available" stationary and mobile source control measures needed to maintain RFP into their revised SIPs.[49] To assist them, the EPA was directed to publish information on procedures and methods to reduce or control pollutants covering the list of potential TCMs contained in Section 108(f) of the act. These "Section 108(f)" TCMs included expanded public transit, high-occupancy vehicle (HOV) lanes, trip reduction ordinances, employer transportation management plans, ride sharing, bicycle lanes, pedestrian paths, and parking restrictions.[50] The EPA insisted that "every effort" be made to integrate the air quality related transportation measures required by the Clean Air Act into the planning and programming procedures administered by the DOT.[51]

The 1977 amendments also enforced consistency between transportation planning and the clean air plans. For instance, Section 176(c) of the Clean Air Act required all federal projects, licenses, permits, financial assistance, and other activities to conform to an approved SIP. The Federal Highway Administration (FHWA) in the DOT was made responsible for assuring the conformity of any federally funded highway projects, though by agreement it would consult with the EPA.[52] In addition to these restrictions on federal actions, the section also prohibited any local metropolitan planning organization (MPO) from approving any "project, program or plan," even those without federal funding, that did not conform. Congress set up MPOs for each urban area of more than 50,000 in population to provide for "continuing, coordinated and comprehensive" transportation planning between federal, state and local authorities, known as the "3C" planning process.[53]

The 1977 amendments did not define *conformity* directly, but the EPA and the DOT jointly issued a guidance document in 1980 for establishing the conformity of transportation plans, improvement programs, and projects with SIPs.[54] The DOT subsequently issued an interim final rule based on the June 1980 joint guidance to cover conformity determinations for federally financed highway projects in nonattainment areas.[55] Under these initial guidelines, transportation plans and programs conformed with the SIP if they did not adversely affect the TCMs in the SIP and if they contributed to reasonable progress in implementing those TCMs.[56] The regulations relied principally on the local MPO to make the initial conformity finding and focused more on ensuring the adequacy of the process for preparing plans and programs than evaluating the impacts from particular highway projects.[57] A project would conform as long as it came from a conforming transportation improvement program or did not otherwise adversely affect the TCMs in the SIP.[58]

Thus, under the 1977 amendments and federal regulations, conformity was interpreted at this time in the context of implementing TCMs rather than performing air quality analyses or measuring emissions levels against RFP targets.[59] In other words, once a region had made an attainment demonstration, it could approve highway projects, consistent with the planned highway network and the schedule for TCM implementation, without conducting a new air quality assessment to demonstrate that the proposed projects were consistent with attaining the NAAQS. The EPA remained concerned, however, that highway projects might be approved that could cause violations of the Clean Air Act in attainment areas or interfere with progress toward meeting the standards in nonattainment areas. Although responsibility for approving highway projects ultimately rests with the DOT, the EPA believed that state SIPs should establish criteria and procedures to help ensure that all federal actions were in conformity to the SIP.[60]

Requirements for Further Extensions to 1987

Not all states could be expected to meet the timetable for compliance. Those states with severe pollution problems, such as the Bay Area, could obtain an extension to December 31, 1987, if, despite implementation of all required control measures, the NAAQS could still not be attained by the original 1982 deadline.[61] To receive the extension, the state's nonattainment plan had to include an I/M program and identify any other measures necessary to attaining the NAAQS by the new date.[62] States also had to revise their SIPs by July 1, 1979, to include comprehensive measures to expand public transit and to implement any TCMs needed to meet the federal standards.[63] A second revision was due by July 1, 1982,[64] in which the states had to commit to adopting "enforceable measures" to ensure timely attainment of the NAAQS.[65]

In January 1981, the EPA adopted regulations that established guidelines for the 1982 plan revisions. The EPA's SIP approval policy called for a "fully adopted, technically justified program," which would result in attainment by 1987. Plans had to contain three categories of minimum control measures: (a) reasonably available control technology (RACT) for stationary sources, (b) a vehicle I/M program, and (c) all "reasonably available" TCMs.[66] If attainment could still not be demonstrated by the end of 1987, additional control measures to be implemented after that date had to be identified in the SIP, and attainment had to be shown as soon as possible afterward.[67]

The plan also had to show that reasonable further progress toward attainment of the ozone and CO standards would continue during any period of nonattainment authorized by the extension. At that time, RFP was defined to mean such "annual incremental reductions in emissions" of the applicable air pollutant as were sufficient, in the judgment of the EPA, to achieve the NAAQS by the applicable attainment date.[68] The RFP demonstration had to indicate the total amount of the annual reductions in emissions and to distinguish between projected reductions from mobile source and stationary source pollution control measures. An annual RFP report demonstrating compliance, including the status of any proposed TCMs, had to be submitted by July 1 of each year.[69]

The transportation portion of the plan had to include a list of all Section 108(f) TCMs to be adopted[70] and a monitoring plan to assess the success or failure of the TCMs in achieving the projected emission reductions.[71] In addition, to satisfy the MPO's obligations under Section 176(c), the EPA regulations provided it had to contain a conformity assessment procedure for ensuring that all approved transportation plans, programs, and projects conformed to the SIP.[72] The EPA's 1981 SIP approval policy also required certain areas of the country to adopt a two-part contingency plan, consisting of

1. a list of planned transportation projects to be delayed, while the SIP is being revised, if expected emissions reductions or air quality improvements did not occur, and
2. a process to select and implement additional TCMs to compensate for any unanticipated shortfalls in emission reductions.

The state had to initiate the contingency provisions when the EPA determined that the SIP was inadequate to attain NAAQS and that additional emissions reductions were needed.[73]

The adequacy of both the Bay Area's conformity assessment procedures and its contingency plan was challenged in the litigation described in the next chapter. The federal district court held that MTC's conformity assessment did not adequately quantify emissions levels from transportation sources, a ruling that forced the agency to develop and apply new modeling procedures that improved on the standard travel demand model discussed earlier. The court also decided that the MTC and other defendants had failed to adopt appropriate contingency measures once it became apparent that the region had not stayed "on the RFP line" in its efforts to reduce pollution. The new travel model became an important tool in evaluating whether additional control measures would generate sufficient emissions reductions to bring the region back into compliance with the Clean Air Act as the court had directed. These issues were affected by the passage of the 1990 amendments and ISTEA, described next, which occurred while the litigation was in progress. Both sides argued that the changes in the law supported their position on critical matters.

Progress Under the 1977 Amendments

Despite the 1977 congressional reforms, states continued to fall behind in their efforts to improve the nation's air quality. A substantial number of areas failed to meet the 1982 deadline for NAAQS compliance.[74] Although overall the 1977 Clean Air Act did achieve a cleaner vehicle fleet, a major problem area was the rapid growth of SOV trips, which increased traffic volumes leading to higher pollution levels.[75] By 1987, it was apparent that many extension areas would also not achieve NAAQS on time and that only draconian measures would be able to remedy the violations in the short term. The EPA recognized, though, that the intent of the 1977 amendments was to achieve cleaner air without undue interference in economic growth. Seeking to accommodate these conflicting goals, the agency undertook efforts to force some states to adopt tougher pollution controls.[76] It also began considering policies to deal with the problem of persistent nonattainment areas in an effort to avoid the need to prepare FIPs for any area not reaching attainment by the end of 1987 despite using all reasonably available measures. At the same time, efforts were already under way in Congress to again amend the Clean Air Act.

For areas such as the Bay Area, which had approved Part D SIPs but were still violating the ozone and CO standards after December 31, 1987, the agency proposed to require a new round of planning and intended to reserve sanctions, including mandatory construction bans and loss of federal funds for transportation projects, only for those areas that failed to submit adequate plan revisions.[77] The EPA planned to notify a number of nonattainment areas by spring 1988 that their SIPs were "substantially inadequate" and would need to be revised. The Bay Area was expected to be one of the areas receiving a "SIP call." To avoid sanctions, the revised implementation plans would have to contain a "persuasive demonstration" that after 1987, new strategies would result in attainment in a short period of time—within 3 to 5 years from EPA approval. Those areas projecting near-term attainment would have to also ensure maintenance of the standards for at least 10 years. Areas that could not meet the new deadlines despite making "reasonable efforts" could have longer before funding sanctions would be applied but would be subject to the construction ban.[78] Noting the historical problems encountered with imposing federal clean air plans, the EPA decided it would prepare a FIP only if the sanctions failed to induce an area to develop a corrective SIP in response to its notice. Congress, the agency reasoned, had intended the threat of a construction ban to encourage states to adopt their own plans and thus avoid EPA involvement in developing local transportation control plans if at all possible.[79]

All except the most marginal nonattainment areas would have to demonstrate a reasonable rate of progress toward attainment by achieving at least a 3% average annual reduction in 1987 base year HC emissions.[80] This *rate of progress* requirement was designed to deal with the fact that many earlier plans had contained overly optimistic estimates of emission reductions and delayed most of the reductions until the later years of the plans so that it was difficult to determine the likelihood of attainment until well into the implementation process. States would have to assess their compliance by the fifth year from the SIP call. Thus, a state's annual progress report for 1992 would have to show a minimum 15% reduction by that year. These areas would have to account for any growth in mobile or stationary source emissions expected to occur between the base year and the attainment date.[81] Long-term CO nonattainment areas with localized hot spot problems would also have to commit to implementing solutions within 3 years, and those with areawide CO problems would have to meet the 3% annual reduction target.[82] The EPA believed that the threatened construction ban and potential loss of federal funds would induce problem areas to improve their pollution control efforts. As discussed in the following chapter, though, the plaintiffs were much less patient and wanted the local agencies to take immediate action to correct all existing deficiencies.

In addition to these policies, the EPA also suggested changes in the 1977 act designed to complement its own administrative actions. Among these were proposals to vary the attainment dates by the relative difficulty of achieving the federal standards. At the same time, the EPA was also urging the DOT to revise its regulations to evaluate the impact of transportation plans and programs on achieving and maintaining NAAQS and to analyze individual highway projects in making their conformity determinations.[83] These policies and legislative suggestions would provide the groundwork for efforts to amend the 1977 act. Although the Reagan administration had initially sought to weaken provisions of the Clean Air Act, Congress responded by actually strengthening it.[84] The bill signed by President Bush on November 15, 1990, made several key changes in the act.

The 1990 Amendments

The latest amendments to the Clean Air Act renewed the federal government's earlier reliance on transportation control measures to combat air pollution that was temporarily sidestepped in the 1974 and 1977 amendments. They also forged new linkages between air quality and transportation planning. The growth in vehicle use, about 3% per year nationwide, was found to be a major factor in states failing to meet the 1987 deadlines. Congress finally recognized that it could not solve the air pollution crisis merely through stationary source controls or by making cars run cleaner. The 1990 amendments were designed to reduce emissions from vehicle use, particularly in major urban centers where pollution was worst. As the Congressional Clean Air Conference Committee Report explains,

> The transportation control provisions of the bill are designed to correct 20 years of failed efforts to control transportation sources of pollution. The sponsors hope that we have learned from the mistakes of the past and have designed a better approach to achieving those objectives.[85]

As recommended by the EPA, the 1990 amendments further extended the deadlines for compliance for nonattainment areas based on the severity of their pollution problems but also mandated that stricter measures be adopted to control air pollution. The amendments endorsed tighter tailpipe standards, advanced I/M programs, fleet conversions to alternative fuels, and stronger links between regional transportation plans and ambient air quality standards. The EPA's discretion in imposing sanctions for delays was eliminated.

Instead, failure to submit an adequate plan can now result in automatic penalties.[86]

The amendments classified HC nonattainment areas as either "Moderate," "Serious," "Severe," or "Extreme," based on their *design value* for the base year. The design value is the projected daily pollution concentration for each pollutant, which is expected to be exceeded only once per year. Similarly, CO nonattainment areas were classified as either "Moderate" or "Serious." Areas with higher classifications were given later attainment dates but are subject to tighter regulations. States that do not meet the standards by the specified dates can be "bumped up" to the next higher classification.[87]

By November 15, 1992, all nonattainment areas had to prepare updated comprehensive emissions inventories.[88] States also had to submit a SIP revision demonstrating attainment of the NAAQS by the applicable attainment date, containing new control strategies for reducing emissions and providing for specific annual reductions in pollutant levels.[89] The plan had to provide for the expeditious implementation of all "reasonably available" TCMs[90] and contained specific contingency measures that could automatically take effect without further action by the state or the EPA if the area failed to make RFP or to meet the applicable deadlines for the NAAQS.[91]

The new law pays special attention to curbing the growth in emissions resulting from any increase in VMT or the number of vehicle trips (VTs) in the region.[92] For those Moderate CO areas with a design value above 12.7 ppm, their SIPs must contain estimates of VMT for each year before the projected attainment year.[93] If actual VMT exceeds the forecast or NAAQS are not achieved, then any contingency TCMs become effective immediately.[94] Stiffer requirements apply to Severe and above ozone areas and Serious CO areas. They must cap automotive emissions at their current levels. By the end of 1992, these areas must have adopted a SIP that contains "specific and enforceable" TCMs and other transportation control strategies to offset any growth in emissions from increased VMT or vehicle trips.[95] Congressman Henry Waxman, a principal author of the bill, has stated,

This vitally important provision is intended to assure that transportation planning will be used to prevent emission increases due to population growth and increased vehicular traffic in polluted areas. Without [this section] pollution reduction benefits resulting from controls on stationary sources and other sectors of the emissions inventory will be eroded by additional pollution from unprojected increases in vehicle miles traveled and congestion.[96]

These new requirements may well finally force ozone and CO nonattainment areas to become serious about reducing traffic levels.[97]

ISTEA

The 1990 Clean Air Act Amendments are complemented by ISTEA.[98] The metropolitan planning provisions of ISTEA make the local MPO responsible, in cooperation with the state and local transit operators, for developing both a long-range regional transportation plan (RTP) and a transportation improvement program (TIP) containing a list of projects consistent with the RTP.[99] The RTP is a *planning document* that contains policies to accommodate current and future travel demands. It includes all proposed long-term projects in the region and covers a minimum 20-year period.[100] The TIP is an *implementation device,* actually a set of project funding priorities for the short term (3 years), which must be consistent with the RTP. It must be updated every 2 years and has to include all local federally funded transportation projects proposed for that time period.[101]

In addition, ISTEA expands planning at the statewide level to require a statewide planning process, a statewide transportation plan, and a statewide transportation program. As part of the planning process, each state must also develop traffic congestion management systems for all areas of more than 200,000 in population, known as transportation management areas or TMAs. All transportation plans and programs in a TMA must be based on the "3C" planning process carried out by the MPO in cooperation with the state and local transit operators.[102] The MPOs in certain designated TMAs must include officials of major local transit agencies along with local and state elected officials.[103] The EPA has recognized that the additional responsibilities placed on MPOs for transportation planning as well as the increased focus on congestion management will require more sophisticated transportation modeling.[104]

Air pollution control is an important part of ISTEA. In TMAs that contain nonattainment areas for ozone and CO, all highway projects that significantly increase SOV capacity must be part of the approved congestion management system.[105] Failure to comply can result in a 10% penalty of apportioned highway and transit funds. On the other hand, additional funds are available in the act's Congestion Mitigation and Air Quality Improvement Program (CMAQ) for transportation projects (including TCMs) that contribute to attaining NAAQS.[106]

Air Quality/Transportation Linkages

Together the 1990 Clean Air Act Amendments and ISTEA encourage air quality improvement through coordination between the federally mandated transportation planning process and air quality planning for the SIP.[107] Each region must quantify its mobile source emissions and must develop its local transportation plans around achieving levels of vehicle use consistent with the targets established for meeting the NAAQS. The new laws support this process by, for example, mandating that the RTPs and TIPs use the same data for current and projected vehicle emissions that the SIPs use.[108] Furthermore, MPOs must give funding priority in their TIPs to all TCMs listed in the SIP to ensure their "timely implementation."[109] As mentioned, ISTEA also provides funding for alternative modes of travel that can reduce pollution.

Most important, all RTPs and TIPs must conform to the SIP.[110] The 1990 amendments established criteria for determining conformity that goes beyond the prior DOT regulations. Conformity to a SIP now means conformity to the plan's purpose of eliminating or reducing the severity and number of violations of the NAAQS and achieving expeditious attainment of those standards. It also means that no project, program, or plan may be approved if it conflicts with the air quality goals of the SIP by causing new violations of NAAQS, by worsening existing violations, or by delaying "timely attainment" of any SIP deadline.[111] Moreover, the expected emissions from RTPs and TIPs must be consistent with the emissions estimates and required emissions reductions in the SIP.[112] The new Clean Air Act amendments, in effect, establish a pollution "budget" for most nonattainment areas. From now on, all transportation plans and programs must stay within that budget. Although it is too early to judge their ultimate impact, according to some experts, this may finally force a shift in state and local transportation policies away from accommodating SOV trips toward supporting alternative "personal mobility services" (such as vanpooling) and lead to an overall "systems approach to transportation facilities and programs."[113]

With the 1990 amendments, the conformity requirements have been shifted from simply implementing TCMs to reconciling emissions estimates from transportation plans and programs with the projections and standards in the SIP. According to the EPA, this integration of transportation and air quality planning is "intended to protect the integrity of the implementation plan by ensuring that its growth projections are not exceeded without additional measures to counterbalance the excess growth."[114] The problem of unexpected growth was one of the main issues in the Bay Area litigation, and the new focus on this question is in marked contrast to the practice under the earlier legislation.

All these changes place new demands on transportation planners to integrate transportation objectives with air quality goals. To accomplish this, transportation planners will need the ability to better analyze and predict changes in travel behavior and vehicle emissions. As the DOT recognizes, though, existing transportation models were not designed to estimate pollution levels. Although these models—which rely on approximations of population, household characteristics and land uses to generate estimates of traffic volume and travel patterns—may be suitable for traffic engineering purposes, they are simply not accurate enough for the required emissions modeling.

Estimating emissions requires data on traffic volume and average speeds at various times of day, neither of which are typically generated in the current transportation models. This information must be obtained and combined with estimates of different pollutants emitted by various types of vehicles. Most such estimates are based on the EPA's MOBILE model, which uses emission rates taken from a sample of vehicles. The data must be refined by correction factors to account for regional driving patterns and air temperatures.[115]

There are also a number of other problems to estimating future vehicle emissions. Better data are needed on travel demand, emissions levels from the current and expected automobile fleets in different areas, and the levels of pollution generated by both stationary sources and vehicles. The relationship between land use and transportation needs to be better understood. Air pollution may also be affected by both atmospheric conditions and traffic patterns. There is also little data on the effectiveness of various TCMs in reducing traffic and therefore emissions.[116] Another frequently expressed concern is the lack of familiarity among transportation planners with the MOBILE emissions model and the EPA's regional air quality models.[117]

Clearly, the new laws will have an impact on transportation planning in the future. The amendments put pressure on transportation planners to adopt TCMs and increase the pressure on any proposed projects that might contribute to sprawl, VMT, or additional congestion.[118] Not only will planners need to take greater account of the air pollution impacts of transportation and land use decisions, but compliance with the legal requirements will also likely stretch the capacity of current planning techniques to quantify and document those effects.

Many of the issues raised by the 1990 amendments and the changes in ISTEA were addressed or at least anticipated by the Bay Area litigation to which we turn next. Both the new legislation, and the interpretation given to the mandates of the Clean Air Act by this court, represent a dramatic shift in the relationship between transportation planning and air quality. They reflect a growing recognition that each airshed region in the country has a carrying capacity for pollutants that cannot be exceeded without jeopardizing the health of the people living there. As a result, Congress and the EPA have concluded that air quality concerns should replace mobility considerations as the overriding factor in highway planning.

This case illustrates the difficulty facing planners in upgrading travel forecasting to meet the challenge posed by the new legislation. Chapter 2 discusses the factual background of the case and the court's initial rulings concerning the scope of the defendants' liability for carrying out their SIP. The major issues to be addressed concern the grounds for holding that the defendant agencies' commitments in the state implementation plan were legally enforceable and, to the extent that they were, whether the defendants had made reasonable progress toward meeting their legal obligations.

Chapters 3 and 4 deal with the specific planning issues raised in the case. Chapter 3 discusses the nature and level of sophistication of the tools the planners were required to use to assess whether future transportation plans and projects conformed to the air quality standards contained in the SIP. Among the disputes between the parties were the proper way to define conformity consistent with the Clean Air Act and EPA regulations, whether conformity assessments could be carried out just for the transportation plans and programs or also had to be done for individual transportation projects, the length of the time period over which conformity must be demonstrated, and the extent to which any potential impacts of highway construction on regional growth and development patterns should be taken into account in these assessments.

Chapter 4 addresses whether the TCMs proposed and adopted by the regional planners, both prior to and as a direct result of this litigation, would produce sufficient reductions in pollution emissions to meet the federal air quality targets. Here, the dispute centered on how much accuracy can be expected in assessing the future performance of individual control measures designed to reduce emissions levels, particularly where there is little or no empirical evidence to back them up. Related to this is the question of how to deal with the fact that some of the data—and even the methodologies used to prepare the initial plans—may become outmoded with time or incompatible with newer data and improved modeling approaches.

Both of the major issues raised in these two chapters, conformity and compliance with emission reduction schedules, illustrated major shortcomings in the 1977 act that were on the minds of the federal regulators when the case was filed. The failure of many areas to achieve their expressed air quality goals, and the ongoing controversy between the DOT and the EPA over the air quality impacts of transportation planning and investment, caused a rethinking of the government's approach to air pollution problems. Because some of the same individuals involved in the Bay Area litigation were also involved in drafting the legislative amendments to the Clean Air Act, the EPA was aware of the significance of the rulings being made in the case and it no doubt had an effect on the legislative process. On the other hand, because the 1990 amendments passed while the case was still in litigation, the final outcome was ultimately affected by the changes made by Congress to the Clean Air Act. As a result, the court specifically addressed the impact of these amendments in its decision. Its rulings thus provide important first insights into how the federal courts may deal with this new legislation.

Chapter 5 summarizes the major holdings of the case and tries to interpret this decision in regard to its potential impact on future transportation planning and modeling practice. The EPA has already begun issuing new regulations governing these matters, and we offer some preliminary analysis. Although other cases are sure to follow as the courts are called on to further interpret the new congressional program in light of actual experiences, this case marks a turning point in our nation's approach to transportation and air quality planning.

Notes

1. Clean Air Act Amendments of 1990, Pub. L. No. 101-549, 104 Stat. 2399 (1990) (codified as amended at 42 U.S.C.A. §§ 7401 et seq. (West 1995)).

2. Intermodal Surface Transportation Efficiency Act of 1991, Pub. L. No. 102-240, Title I, 105 Stat. 1914 (1991) (codified as amended at 23 U.S.C.A. §§ 101 et seq. (West 1990 & Supp. 1995)).

3. See, for example, 23 U.S.C.A. §§ 134(f)(4), 135(c)(14) (West Supp. 1995).

4. 42 U.S.C.A. § 7401(2) (West 1995).

5. U.S. Department of Transportation, *Intermodal Surface Transportation Efficiency Act of 1991: A Summary,* not dated.

6. P. R. Stopher and A. H. Meyburg, *Urban Transportation Modeling and Planning* (Lexington, MA: D. C. Heath, 1975), chapter 1.

7. Id. Chapters 11 and 12.

8. G. Harvey and E. Deakin, *A Manual of Regional Transportation Modeling Practice for Air Quality Analysis* (National Association of Regional Councils, July 1993).

9. Clean Air Act, Pub. L. No. 159, ch. 360, 69 Stat. 322 (1955), *as amended by* Pub. L. No. 86-365, 73 Stat. 646 (1959), Pub. L. No. 88-206, 77 Stat. 392 (1963), *renumbered and amended by* Pub. L. No. 89-272, Title I, § 101, 79 Stat. 992 (1965) (originally codified at 42 U.S.C. §§ 1857 et seq.).

10. Air Quality Act of 1967, Pub. L. No. 90-148, § 2, 81 Stat. 485 (1967).

11. Clean Air Amendments of 1970, Pub. L. 91-604, 84 Stat. 1676 (1970).

12. Pub. L. No. 91-604, § 4(a), 84 Stat. 1679 (1970) (adding § 109 to the Clean Air Act, codified at 42 U.S.C.A. § 7409(a) (West 1995)) There are also limits set for nitrogen dioxide, sulfur dioxide, particulate matter, and lead

13. Nitric oxide (NO) and nitrogen dioxide (NO_2) are together referred to as oxides of nitrogen (NOx). Oxides of nitrogen are emitted primarily as NO and transform rapidly via chemical reactions with available oxygen in the atmosphere into NO_2.

14. Generally, several hours are required for reactions of precursors to reach maximum level, creating the highest ozone concentrations downwind of the source of precursor emissions, and several hours later in the day. Therefore, emissions of precursors in the morning typically lead to elevated ozone concentrations in the afternoon, and early morning emissions play a more important role than late day primary pollutant emissions in determining ozone levels for a given day.

15. Pub. L. No. 91-604, § 4(a), 84 Stat. 1680 (1970) (adding § 109(b)(1) to the Clean Air Act, codified at 42 U.S.C.A. § 7409(b)(1) (West 1995)). The act also established national secondary standards, which are the levels of air quality necessary to protect the public welfare from any known or anticipated adverse effects of a pollutant. Id. at subsection (b)(2).

16. 40 C.F.R. § 50. 9 (1994). The number of daily exceedences is averaged over a 3-year period.

17. 40 C.F.R. § 50. 8 (1994).

18. Pub. L. No. 91-604, § 4(a), 84 Stat. 1678 (1970) (adding § 110(a)(1) to the Clean Air Act), *amended by* Pub. L. 101-549, Title I, § 101(c)(8), 104 Stat. 2408 (1990) (current version at 42 U.S.C.A. § 7410(a)(1) (West 1995)). The section originally required that the SIP be adopted and submitted to the EPA administrator within 9 months after the promulgation of the NAAQS or any subsequent revisions. The EPA was authorized to designate as an air quality control region any interstate area or major intrastate area deemed necessary or appropriate for attainment and maintenance of air quality standards. Pub. L. No. 91-604, § 4(a), 84 Stat. 1678 (1970) (adding § 107(c) to the Clean Air Act), *amended by* Pub. L. 95-95, § 103, 91 Stat. 687 (1977) (current version at 42 U.S.C.A. § 7407(c) (West 1995)).

19. Pub. L. No. 91-604, § 4(a), 84 Stat. 1678 (1970) (adding § 110(a)(2)(B) to the Clean Air Act, originally codified at 42 U.S.C. § 7410(a)(2)(B)), *amended by* Pub. L. 95-95, § 108(a)(2),

91 Stat. 693 (1977); Pub. L. 101-549, § 101(b), 104 Stat 2404 (1990) (see now 42 U.S.C.A. § 7410(a)(2)(A) (West 1995)).

20. Pub. L. No. 91-604, § 4(a), 84 Stat. 1678 (1970) (adding § 110(a)(2)(A)(i) to the Clean Air Act, originally codified at 42 U.S.C. § 7410(a)(2)(A)(i)), *amended by* Pub. L. 95-95, Title I, § 108(a)(1), 91 Stat. 693 (1977); Pub. L. 101-549, Title I, § 101(b), 104 Stat. 2402 (1990). This section provided that the NAAQS be achieved "as expeditiously as practicable" but no later than 3 years from the date the plan was submitted to and approved by the EPA. An extension of up to 2 years could be granted by the EPA under specified conditions. Id. (adding § 110(e), originally codified at 42 U.S.C. § 7410(e)), *repealed by* Pub. L. No. 101-549, Title I, § 101(d)(5), 104 Stat. 2409 (1990).

21. Pub. L. No. 91-604, § 4(a), 84 Stat. 1680 (1970) (adding § 110(a)(2)(G) to the Clean Air Act, originally codified at 42 U.S.C. § 7410(a)(2)(G)), *amended by* Pub. L. No. 101-549, § 101(b), Title IV, § 412, 104 Stat. 2404 (1990).

22. Pub. L. No. 91-604, § 4(a), 84 Stat. 1678 (1970) (adding § 110(a)(2)(H) to the Clean Air Act), *amended by* Pub. L. 95-95, § 108(a)(6), 91 Stat. 693 (1977); Pub. L. 101-549, § 101(b), 104 Stat. 2404 (1990) (current version at 42 U.S.C.A. § 7410(a)(2)(H) (West 1995)).

23. Pub. L. No. 91-604, § 4(a), 84 Stat. 1678 (1970) (adding § 110(c) to the Clean Air Act), *amended by* Pub. L. 95-95, § 108(d)(1)&(2), 91 Stat. 694 (1977); Pub. L. 101-549, § 102(h), 104 Stat. 2422 (1990) (current version at 42 U.S.C.A. § 7410(c)(1) (West 1995)). Section 110(c)(3) provided that the EPA administrator could issue air quality regulations for a state if the state failed to comply with a notice to revise its SIP pursuant Section 110(a)(2)(H) (see now 42 U.S.C.A. § 7410(c)(1)(C) (West 1995)). See note 22 supra.

24. In *National Resources Defense Council, Inc. v. EPA,* 475 F.2d 968 (D. C. Cir. 1973), the U.S. District Court for the District of Columbia ruled that the Clean Air Act did not permit the EPA to grant states a delay in submission of transportation plans until February 15, 1977 (see 36 Fed. Reg. 15,486) or to grant 2-year extension under Section 110(e) for attaining the primary federal clean air standards where such plans had not been submitted. It also ordered the EPA to inform any states that had not then submitted a complete SIP fully complying with the Clean Air Act to submit such a plan by April 15, 1973, and for the EPA to approve or disapprove the plans by June 15, 1973, and promulgate a plan or portion thereof for any noncomplying state. Sixteen states and the District of Columbia submitted plans that were reviewed by the EPA. See 38 Fed. Reg. 30,626 (November 6, 1973). See also M. T. Donellan, "Transportation Control Plans Under the 1990 Clean Air Act as a Means for Reducing Carbon Dioxide Emissions," *Vermont Law Review, 16*(Winter 1992): 736-738; J. B. Battle, "Transportation Controls Under the Clean Air Act—An Experience in (Un)Cooperative Federalism," *Land and Water Law Review, 15*(1980): 10-12.

25. 38 Fed. Reg. 30,626, 30,629-30 (November 6, 1973). See Donellan, supra note 24, at 729-732; Battle, supra note 24, at 13-16.

26. In the *National Resources Defense Council* case (see note 24 supra) the court also ordered the EPA to review the maintenance provisions of all previously approved SIPs and disapprove those that did not provide for measures to ensure the maintenance of the primary NAAQS after the May 31, 1975, deadline for attaining the federal standards, or did not analyze the problem of maintenance in accordance with the EPA regulations (see former 40 C.F.R. § 51.12(a)). In disapproving those plans that failed to adequately ensure maintenance, the EPA determined that current stationary source controls and tailpipe standards would not be adequate to maintain NAAQS, particularly for pollutants emitted by motor vehicles due to increased use resulting from general urban and commercial development. In an advance notice of proposed rule making, the EPA proposed extending its stationary source review regulations to include pollution from mobile sources associated with constructing certain facilities and to issue

regulations governing procedures for such indirect source review (ISR). See 38 Fed. Reg. 6290 (March 8, 1973). In the preamble to the proposed amendments to 40 C.F.R. Part 51, the EPA recognized the importance of the growth in controlling regional air pollution and encouraged the states to analyze the effects of growth in population and vehicle use on air quality in developing their clean air plans. See 38 Fed. Reg. 9599 (April 18, 1973). The EPA's final published regulations were stronger, noting that because the EPA felt that preconstruction review of individual sources alone could not adequately deal with generalized growth and its impact on regional air quality, the regulations would also require states to perform an analysis of the air quality implications of growth and development, including any increased air pollution arising from, among other things, increases in "motor vehicle traffic." Maintenance of National Ambient Air Quality Standards, 38 Fed. Reg. 15,834 (June 18, 1973) (codified at former 40 C.F.R. § 51.12(g)). In its order, the district court also required the EPA to promulgate implementation plans for any states failing to submit adequate ISR rules by August 15, 1973, in response to the EPA rule making. See 39 Fed. Reg. 7270 (February 25, 1974) See also P. E. Rothchild, "The Clean Air Act and Indirect Source Review: 1970-1991," *UCLA Journal of Environmental Law & Policy* 10 (Winter 1992): 340-344.

27. See, for example, *South Terminal Corp. v. EPA,* 504 F.2d 646 (1st Cir. 1974); *Commonwealth of Pennsylvania v. EPA,* 500 F.2d 246 (3d. Cir. 1974); *State of Maryland v. EPA,* 530 F.2d 215 (4th Cir. 1975), *vacated,* 431 U. S. 99 (1977); *Brown v. EPA,* 521 F.2d 827 (9th Cir. 1975), *vacated,* 431 U. S. 99 (1977); *District of Columbia v. Train,* 521 F.2d 971 (D. C. Cir. 1975), *vacated,* 431 U. S. 99 (1977); *Friends of the Earth v. Carey,* 9 Environment Review Cases (BNA) 1641 (2d Cir. 1977). See also Battle, supra note 24, at 16-24.

28. Donellan, supra note 24 at 732-737; Battle, supra note 24, at 24-30; Rothchild, supra note 26, at 345-347.

29. Energy Supply and Environmental Coordination Act of 1974, Pub. L. No. 93-319, § 4(b), 88 Stat. 257 (1974) (adding § 110(c)(2)(B) to the Clean Air Act, codified at 42 U.S.C.A. § 7410(c)(2)(B) (West 1995)).

30. Pub. L. No. 93-319, § 4(b), 88 Stat. 256 (1974) (adding § 110(c)(2)(E) to the Clean Air Act, codified at 42 U.S.C.A. § 7410(c)(2)(E) (West 1995)).

31. Clean Air Act Amendments of 1977, Pub. L. No. 95-95, 91 Stat. 685 (1977). See General Preamble for Proposed Rulemaking on Approval of Plan Revisions for Nonattainment Areas, 44 Fed. Reg. 20,372 (April 4, 1979). The EPA had previously issued an interpretive ruling allowing new construction in areas where NAAQS were violated as long as stringent conditions were met. 41 Fed. Reg. 55,524 (December 21, 1976). See 40 CFR Part 51, Appendix S, *as revised* 44 Fed. Reg. 3274 (January 16, 1979). The rule prohibited any new construction after January 1, 1979, without an approved SIP revision providing for an "emission offset" exceeding the additional emissions from the new source so as to achieve RFP.

32. Pub. L. No. 95-95, Title I, § 108(d)(3), 91 Stat. 694 (1977) (adding § 110(c)(4) to the Clean Air Act, originally codified at 42 U.S.C. § 7410(c)(4)), *repealed* by Pub. L. No. 101-549, Title I, § 101(d)(3)(C), 104 Stat. 2409 (1990). The suspension would operate to January 1, 1979, by which time the state had to submit a new SIP meeting all the requirements of Part D of the Clean Air Act.

33. Pub. L. No. 95-95, Title I, § 108(d)(3), 91 Stat. 694 (1977) (adding § 110(c)(5)(A) to the Clean Air Act, codified at 42 U.S.C.A. § 7410(c)(5)(A) (West 1995)). States had to submit a SIP revision that included comprehensive measures to expand public transit and implement transportation control measures, which would provide equivalent emissions reductions to that achieved through the eliminated tolls or charges. Id. subsection (c)(5)(B), *amended by* Pub. L. No. 101-549, Title I, § 101(d)(3)(D), 104 Stat. 2409 (1990) (current version at 42 U.S.C.A. § 7410(c)(5)(B) (West 1995)).

34. Pub. L. No. 95-95, Title I, § 108(a), 91 Stat. 693 (1977) (amending § 110(a)(2)(B) of the Clean Air Act, codified at 42 U.S.C. § 7410(a)(2)(B)), *amended by* Pub. L. No. 101-549, Title I, § 101(b), 104 Stat. 2404 (1990). The language replaced the phrase *land use* controls in the earlier version.

35. Pub. L. No. 95-95, Title I, § 108(e), 91 Stat. 695 (1977) (adding § 110(a)(5)(A)(i), codified at 42 U.S.C.A. § 7410(a)(5)(A)(i) (West 1995)). States could still voluntarily include indirect source review (ISR) programs as part of their SIP. The term *indirect source* includes any *road* or *highway* that could attract mobile sources of pollution. Id. (adding § 110(a)(5)(C), codified at 42 U.S.C.A. § 7410(a)(5)(C) (West 1995)). See Rothchild, supra note 26, at 348 (author argues that the 1990 amendments revive ISR-like requirements in the revised list of required TCMs in subsection (1)(A)(xiv) of Section 108(f)).

36. Pub. L. No. 95-95, Title I, § 108(b), 91 Stat. 694 (1977), *as amended by* Pub. L. No. 95-190, § 14(a)(1)-(6), 91 Stat. 1399 (1977) (adding and amending § 110(a)(2)(I) of the Clean Air Act, originally codified at 42 U.S.C. § 7410(a)(2)(I)), *amended by* Pub. L. 101-549, Title I, § 101(b), 104 Stat. 2404 (1990). Section 110(a)(2)(I) originally required each SIP to prohibit the construction of any major stationary source of pollution after June 30, 1979, if emissions would cause or contribute to violating federal standards, unless the plan met the requirements of Part D of the act.

37. Pub. L. No. 95-95, Title I, § 103, 91 Stat. 687 (adding § 107(d) to the Clean Air Act), *amended by* Pub. L. 101-549, § 102(a), 104 Stat. 2412 (1990) (current version at 42 U.S.C.A. § 7407(d) (West 1995)).

38. Pub. L. No. 95-95, Title I, § 129(b), 91 Stat. 745 (1977) (adding Section 171(2) to the Clean Air Act), *amended by* Pub. L. 101-549, § 101(a), 104 Stat. 2399 (1990) (current version at 42 U.S.C.A. § 7501(2) (West 1995)). See 44 Fed. Reg. 20,372, 20,377, ¶ III. A. 1.

39. Pub. L. No. 95-95, Title I, § 129(c), 91 Stat. 750 (1977), *as amended by* Pub. L. 95-190, § 14(b)(4), 91 Stat. 1405 (1977) (adding and amending Section 129(c) of the Clean Air Act).

40. Pub. L. No. 95-95, Title I, § 129(b), 91 Stat. 745 (1977) (adding § 172(a)(1) to the Clean Air Act, originally codified at 42 U.S.C. § 7502(a)(1)), *amended by* Pub. L. No. 101-549, Title I, § 102(b), 104 Stat. 2412 (1990).

41. Pub. L. No. 95-95, Title I, § 129(b), 91 Stat. 745 (1977) (adding § 172(b)(4) to the Clean Air Act, originally codified at 42 U.S.C. § 7502(b)(4)), *amended by* Pub. L. No. 101-549, Title I, § 102(b), 104 Stat. 2412 (1990) (see now 42 U.S.C.A. § 7502(c)(3) (West 1995)). See 44 Fed. Reg. 20,375, ¶ III. A. 1; 43 Fed. Reg. 21,674-5, ¶ 2, 7-8 (May 19, 1978).

42. Pub. L. No. 95-95, Title I, § 129(b), 91 Stat. 745 (1977) (adding § 172(b)(5) to the Clean Air Act, originally codified at 42 U.S.C. § 7502(b)(5)), *amended by* Pub. L. No. 101-549, Title I, § 102(b), 104 Stat. 2412 (1990) (see now 42 U.S.C.A. § 7502(c)(4) (West 1995)). See 44 Fed. Reg. 20,372, 20,375, ¶ III. A. 1 (April 4, 1979); 43 Fed. Reg. 21,673, 21,673, 21,674-5, ¶ 7 (May 19, 1978) ("The growth rates established by states for mobile sources . . . should also be specified, and in combination with the growth associated with major new or modified stationary sources will be accepted so long as they do not jeopardize the reasonable further progress test and attainment by the prescribed date.").

43. Pub. L. No. 95-95, Title I, § 129(b), 91 Stat. 745 (1977) (adding § 172(b)(8) to the Clean Air Act, originally codified at 42 U.S.C. § 7502(b)(8)), *amended by* Pub. L. No. 101-549, Title I, § 102(b), 104 Stat. 2412 (1990) (see now 42 U.S.C.A. § 7502(c)(6) (West 1995)).

44. Pub. L. No. 95-95, Title I, § 129(b), 91 Stat. 745 (1977) (adding § 172(b)(3) to the Clean Air Act, originally codified at 42 U.S.C. § 7502(b)(3)), *amended by* Pub. L. No. 101-549, Title I, § 102(b), 104 Stat. 2412 (1990) (see now 42 U.S.C.A. § 7502(c)(2) (West 1995)). See 44 Fed. Reg. 20,375, ¶ III. A. 1; Criteria for Proposing Approval of Revision to Plans for Nonattainment Areas, 43 Fed. Reg. 21,673, 21,675, ¶ 6 (May 19, 1978).

45. Pub. L. No. 95-95, Title I, § 129(b), 91 Stat. 745 (1977) (adding § 172(b)(6) to the Clean Air Act, originally codified at 42 U.S.C. § 7502(b)(6)), *amended by* Pub. L. No. 101-549, Title I, § 102(b), 104 Stat. 2412 (1990) (see now 42 U.S.C.A. § 7502(c)(5) (West 1995)).

46. Pub. L. No. 95-95, Title I, § 129(b), 91 Stat. 748 (1977), *as amended by* Pub. L. 95-190, § 14(a)(57), 91 Stat. 1403 (1977) (adding and amending § 173(1) of the Clean Air Act), *amended by* Pub. L. 101-549, Title I, § 102(c), 104 Stat. 2415 (1990) (current version at 42 U.S.C.A. § 7503(a)(1) (West 1995)). See 44 Fed. Reg. 20,378, ¶ III. B. 10. a. The EPA's Emissions Offset Interpretive Ruling (see note 31 supra) would terminate after July 1, 1979, and be replaced by either (a) the preconstruction review provisions in the revised SIP, or (b) a prohibition on construction under the applicable SIP and Section 110(a)(2)(I) for SIPs not meeting the Part D requirements. Pub. L. No. 95-95, Title I, § 129(a), 91 Stat. 745 (1977), *as amended by* Pub. L. 95-190, § 14(b)(2),(3), 91 Stat. 1404 (1977). The ruling remained in effect for those situations where it was not superseded (see 44 Fed. Reg. 3275).

47. 40 C.F.R. § 52. 24(a) (1994) (providing that the construction ban authorized by Section 110(a)(2)(I) would operate in any nonattainment area that did not have an approved Part D SIP after June 30, 1979); 44 Fed. Reg 38,471, 38,473 (July 2, 1979). If a state attempted to issue a permit, the EPA could issue an order prohibiting it under Section 113(a)(5). See Pub. L. No. 95-95, Title I, § 111(a), 91 Stat. 704 (1977) (adding and amending § 113(a)(5) of the Clean Air Act), *amended by* Pub. L. 101-549, Title VII, § 701, 104 Stat. 2672 (1990) (current version at 42 U.S.C.A. § 7413(a)(5) (West 1995)). For areas with approved Part D SIPs, Section 173(4) prohibited any permit to be issued for a major new source of pollution if the applicable implementation plan for the state was not being "carried out." 40 C.F.R. § 52. 24(b). See Pub. L. No. 95-95, Title I, § 129(b), 91 Stat. 748 (1977), *as amended by* Pub. L. 95-190, § 14(a)(58), 91 Stat. 1403 (1977) (adding and amending § 173(4) of the Clean Air Act), *amended by* Pub. L. 101-549, Title I, § 102(c), 104 Stat. 2415 (1990) (current version at 42 U.S.C.A. § 7503(a)(4) (West 1995)).

48. Pub. L. No. 95-95, Title I, § 129(b), 91 Stat. 745 (1977), *as amended by* Pub. L. No. 95-190, § 14(a)(59), 91 Stat. 1403 (1977) (adding and amending subsections (a) & (b) of § 176 of the Clean Air Act, originally codified at 42 U.S.C. § 7506(a)&(b)), *repealed by* Pub. L. No. 101-549, Title I, § 110(4), 104 Stat. 2470 (1990). Section 176(a) prohibited the EPA from awarding any grants under the Clean Air Act, and the DOT from approving any highway projects (other than for safety, mass transit, or air pollution control) in any nonattainment region where transportation control measures were necessary for attainment and the state failed to submit an adequate plan or to make reasonable progress toward submitting one. See Final Policy—Federal Assistance Limitations Required by Section 176(a) of the Clean Air Act, 45 Fed. Reg. 24,692 (April 10, 1980). Section 176(b) prohibited the EPA from making any grants to any area where the state or local government failed to carry out its SIP.

49. Pub. L. No. 95-95, Title I, § 129(b), 91 Stat. 745 (1977) (adding § 172(b)(2) to the Clean Air Act, originally codified at 42 U.S.C. § 7502(b)(2)), *amended by* Pub. L. No. 101-549, Title I, § 102(b), 104 Stat, 2412 (1990) (see now 42 U.S.C.A. § 7502(c)(1) (West 1995)). See 44 Fed. Reg. 20,372, 20,375, ¶ III. A. 1 (April 4, 1979); 43 Fed. Reg. 21,673, 21,674-5, ¶ 4 (May 19, 1978).

50. Pub. L. No. 95-95, Title I, § 105, 91 Stat. 689 (1977) (adding § 108(f) to the Clean Air Act), *amended by* Pub. L. 101-549, Title I, § 108(b), 104 Stat. 2466 (1990) (current version at 42 U.S.C.A. § 7408(f) (West 1995)). EPA policy in 1979 was that all Section 108(f) control measures were presumed to be "reasonably available." 44 Fed. Reg. 20,372, 20,377, ¶ III. B. 5 (April 4, 1979); 43 Fed. Reg. 21,673, 21,677 (May 19, 1978) ("[D]ecisions not to implement measures will have to be carefully reviewed to avoid broad rejections of measures based on conclusory assertions of infeasibility."). See *Delaney v. EPA*, 898 F.2d 687, 692 (9th Cir. 1990)

(Court reversed EPA approval of SIP revision after state failed to meet 1982 deadline or obtain an extension because not all "reasonably available" control measures had been included). That policy is no longer in effect. See discussion infra, note 90.

51. 43 Fed. Reg. 21,673, 21,677 (May 19, 1978). EPA realized that this could be a complicated and lengthy process, especially in areas with severe ozone or CO problems, but required that TCMs be initiated before December 31, 1982, even if they could not be fully implemented by that time.

52. Pub. L. No. 95-95, Title I, § 129(b), 91 Stat. 749 (1977), as amended by Pub. L. No. 95-190, § 14(a)(59), 91 Stat. 1403 (1977) (adding and amending § 176(c) of the Clean Air Act), amended by Pub. L. 101-549, § 101(f), 104 Stat. 2409 (1990) (current version at 42 U.S.C.A. § 7506(c) (West 1995)). The section made the heads of the relevant federal department or agency affirmatively responsible for assuring conformity. The Federal-Aid Highway Act of 1970 had directed the Secretary of Transportation to issue guidelines to ensure that future highway construction would be "consistent with any approved plan for the implementation of any ambient air quality standard for any air quality control region." 23 U.S.C.A. 109(j) (West 1990). The EPA and DOT issued a joint guidance, in 1975, which required regional transportation plans and projects to be measured against ambient air standards. The "conformity" provisions in the 1977 Clean Air Act were designed to give legislative authority for the review criteria in the DOT/EPA 1975 joint guidance. 136 Cong. Rec. S16972 (October 27, 1990).

53. Federal-Aid Highway Act, Pub. L. No. 87-866, § 9(a), 76 Stat. 1148 (1962), (adding § 134 to Title 23 U.S.C., ch. 1), amended by Pub. L. No. 91-605, § 143, 84 Stat. 1737 (1970); Pub. L. No. 95-599, § 169, 92 Stat. 2723 (1978); Pub. L. No. 102-240, Title I, § 1024(a), 105 Stat. 1955 (1991); Pub. L. No. 102-388, Title V, § 502(b), 106 Stat. 1566 (1992); Pub. L. 103-429, § 3(5), 108 Stat. 4377 (1994) (current version at 23 U.S.C.A. § 134 (West Supp. 1995)). As currently written, this section provides that it is in the national interest to "encourage and promote the development of transportation systems to maximize the mobility of people and goods within and through urbanized areas and minimize transportation-related fuel consumption and air pollution" 23 U.S.C.A. § 134(a). To carry out the mandated planning process, MPOs are established in each urbanized area of more than 50,000 population. Id. subsection (b). The boundaries are designated by agreement between the MPO and the governor of the state but they must cover the existing urbanized area and any contiguous areas expected to become urbanized within 20 years. If the area is a designated nonattainment area for CO or ozone, then the boundaries of the MPO must at least include the nonattainment area. Id. subsection (c).

54. Procedures for Conformance of Transportation Plans, Programs and Projects with Clean Air Act State Implementation Plans (June 12, 1980). This document grew out of a June 1978 Memorandum of Understanding Regarding Integration of Transportation and Air Quality Planning between the DOT and the EPA, dealing with conformity procedures for programs administered by FHWA and the Urban Mass Transportation Administration (UMTA), which allowed EPA to review and comment on the conformity of transportation plans and TIPs.

55. Air Quality Conformity and Priority Procedures for Use in Federal-Aid Highway and Federally Funded Transit Programs, 46 Fed. Reg. 8429 (January 26, 1981) (revising 23 C.F.R. Part 770—FHWA Air Quality Guidelines, removed by 57 Fed. Reg. 60,728 (December 22, 1992)). The 1980 agreement and the regulations provided that the conformance findings meeting these standards would satisfy the consistency requirements under 23 U.S.C. § 109(j), thus superseding the 1975 joint guidelines. Id. (former 23 C.F.R. § 770. 7).

56. Id. (former 23 C.F.R. § 770. 9(a)).

57. The 1980 agreement and subsequent DOT regulations provided that the FHWA and UMTA (now the Federal Transit Administration) would jointly determine whether the TIP

conformed with the SIP based in part on (a) the MPO's determination of conformity, (b) a finding by FHWA and UMTA that the urban transportation process effectively incorporated air quality objectives and procedures in the development of the plan and program, (c) that coordination existed between air quality and transportation agencies, and (d) that there had been timely programming and implementation of the TCMs contained in the SIP. Id. (former 23 C.F.R. § 770. 9(b)(1)). In the event of a finding that reasonable progress was not being made on the commitments in the SIP, the FHWA and UMTA would consult with the local agencies and the EPA before making a final conformity determination on accelerating implementation of TCMs and alternatives for delayed projects. Id. at 8430 (former § 770. 9(b)(2)). If a conformity finding could not be made, the FHWA would restrict further construction funding. Id. (former § 770. 9(b)(3)). See also U.S. General Accounting Office, *Air Pollution: EPA Needs More Data From FHWA on Changes to Highway Projects,* March 1990, Appendix V.

58. 46 Fed. Reg. 8430 (former 23 C.F.R. § 770. 9(c)). After approval of the final environmental impact statement (EIS) or a finding of no significant impact, the project would not be subject to any further conformity review unless a supplemental EIS significantly related to air quality considerations was undertaken, a SIP revision was requested, or major steps toward implementation of the project had not commenced within 3 years of the date of approval of the final EIS. Id. (former § 770. 9(d)).

59. Criteria and Procedures for Determining Conformity to State or Federal Implementation Plans of Transportation Plans, Programs, and Projects Funded or Approved Under Title 23 U.S.C. or the Federal Transit Act, 58 Fed. Reg. 62,188, 62,189 (November 24, 1993) [hereinafter Conformity Rule].

60. Conformity of Federal Actions to State Implementation Plans, 45 Fed. Reg. 21,590 (April 1, 1980). In this advance notice of proposed ruling, the EPA suggested that conformity to a SIP should include a requirement of consistency with the state's approach for demonstrating reasonable further progress during the period prior to attainment of the NAAQS and that any increased emissions from all covered federal actions must be accommodated in the emissions growth increment in the SIP.

61. Pub. L. No. 95-95, Title I, § 129(b), 91 Stat. 745 (1977) (adding § 172(a)(2) to the Clean Air Act, originally codified at 42 U.S.C. § 7502(a)(2)), *amended by* Pub. L. No. 101-549, Title I, § 102(b), 104 Stat. 2412 (1990). See 44 Fed. Reg. 20,372, 20,375, ¶ III. A. 2 (April 4, 1979).

62. Pub. L. No. 95-95, Title I, § 129(b), 91 Stat. 745 (1977) (adding § 172(b)(11)(B)&(C), originally codified at 42 U.S.C. § 7502(b)(11)(B)&(C)), *amended by* Pub. L. No. 101-549, Title I, § 102(b), 104 Stat. 2412 (1990). See 44 Fed. Reg. 20,375, ¶ III. A. 2. These included measures that may not have been reasonably available in 1979 but were needed to meet the 1987 deadline. See State Implementation Plans for Nonattainment Areas for Ozone and Carbon Monoxide, 52 Fed. Reg. 26,404, 26,407 (July 14, 1987).

63. Pub. L. No. 95-190, § 14(a)(4), 91 Stat. 1399 (1977) (adding § 110(a)(3)(D), originally codified at 42 U.S.C. § 7410(a)(3)(D)), *repealed by* Pub. L. No. 101-549, Title I, § 101(d)(1), 104 Stat. 2409 (1990). See 44 Fed. Reg. 20,375, ¶ III. A. 2; 43 Fed. Reg. 21,675-6, ¶¶ 2-3. Prior to repeal, this section required that the applicable implementation plan include the comprehensive measures identified in Section 110(c)(5)(B) (see discussion in note 33 supra).

64. Pub. L. No. 95-95, Title I, § 129(c), 91 Stat. 750 (1977), *as amended by* Pub. L. No. 95-190, § 14(b)(4), 91 Stat. 1405 (1977) (adding and amending Section 129(c) of the Clean Air Act). Section 129(c) provided in part that the plan revision had to meet all the requirements of Section 172(b) and (c) of the Clean Air Act listing the required nonattainment plan provisions.

65. Pub. L. No. 95-95, Title I § 129(b), 91 Stat. 746 (1977) as amended by Pub. L. 95-190, § 14(a)(56), 91 Stat. 1402 (1977) (adding and amending § 172(c) to the Clean Air Act, originally

codified at 42 U.S.C. § 7502(c)), *amended by* Pub. L. No. 101-549, Title I, § 102(b), 104 Stat. 2412 (1990) (see now 42 U.S.C.A. § 7502(c)(6) (West 1995)). These included those measures identified in the previous 1979 SIP submittal.

66. Approval of 1982 Ozone and Carbon Monoxide Plan Revisions for Areas Needing an Attainment Date Extension, 46 Fed. Reg. 7182, 7185-6, ¶¶ I. A-D (January 22, 1991); see also 44 Fed. Reg. 20,375, ¶ III. A. 2.

67. Criteria for Approval of the 1982 Plan Revisions, 46 Fed. Reg. 7182 7188, ¶ I. F. (January 22, 1981). The EPA also required more extensive evidence as to why any of the Section 108(f) TCMs were not included in the plan.

68. Pub. L. No. 95-95, Title I, § 129(b), 91 Stat. 746 (adding § 171(1)), *amended by* Pub. L. No. 101-549, Title I, § 102(b), 104 Stat. 2412 (1990) (current version at 42 U.S.C.A. § 7501(1) (West 1995)). The EPA regulations defined RFP to mean that by the end of 1982, annual emission reductions must at least equal the emission reductions represented graphically by a straight line drawn from the emissions inventory for the 1979 base year to the allowable emissions on the attainment date. 46 Fed. Reg. 7182, 7187-8, ¶ I. E. (January 22, 1981). See also, 44 Fed. Reg. 20,372, 20,376-7, ¶ III. B. 3 (April 4, 1979); 43 Fed. Reg. 21,673, 21,675, ¶ 6 May 19, 1978).

69. 46 Fed. Reg. 7182, 7187-8, ¶ I. E. (January 22, 1981). Section 172(b)(4) of the 1977 Clean Air Act authorized the EPA to require annual reports documenting reductions in emissions for CO and ozone to ensure RFP and attainment (see now 42 U.S.C.A. § 7502(c)(3) (West 1995)). The RFP reports were to be submitted as part of the annual source emissions and state action report required by 40 C.F.R. §§ 51. 321-51. 328 (1994).

70. 46 Fed. Reg. 7182, 7187, ¶ I.D. 2. (January 22, 1981). The plan had to present documentation, based on technical analysis, of the basis for not implementing any of the Section 108(f) TCMs. See also 44 Fed. Reg. 20,377 ("[T]he burden is on the state or local government to demonstrate the unavailability of the measure, based on the local situation. A demonstration that a measure is not reasonably available must be based on substantial widespread and long-term adverse impact that would result from the measure, and on the time needed to analyze, develop and implement the measure.")

71. 46 Fed. Reg. 7187, ¶ I.D. 6.

72. Id. at 7188, ¶ I. G. The 1981 SIP approval policy called for the plan to identify the emissions associated with all major federal actions to facilitate state and local review of conformity determinations. The EPA intended to clarify its conformity policy in a final rule based on its April 1, 1980, advance notice of proposed rule making, but no further action was taken. See 45 Fed. Reg. 21,590.

73. 46 Fed. Reg. 7182, 7187, ¶ I.D. 8 (January 22, 1981). All areas had to submit a description of the process to be used to implement additional transportation measures to compensate for unanticipated shortfalls in emission reductions. The procedures for delaying projects were applicable only to areas of more than 200,000 in population. These areas had to include a locally developed list of projects that the implementing agencies had agreed could be delayed during an interim period while the SIP was being revised. For a 12-month period, UMTA and FHWA would not authorize construction of any project on the list unless it was exempt from sanctions under 42 U.S.C. § 7506(a). See former 23 C.F.R. § 770. 9(e)(2) (originally published at 46 Fed. Reg. 8430), *removed by* 57 Fed. Reg. 60,728 (December 22, 1992). Following the December 31, 1987, deadline, the DOT proposed to remove this restriction on funding highway projects during a SIP revision. See Air Quality Procedures for Use in Federal-Aid Highway and Federally Funded Transit Programs, 53 Fed. Reg. 35,178, 35,182 (September 9, 1988).

74. The EPA initially intended to impose the construction ban in Section 110(a)(2)(I) of the Clean Air Act on any nonextension areas that failed to attain the primary NAAQS by December

31, 1982, even if it had previously approved the state's SIP. Compliance with the Statutory Provisions of Part D of the Clean Air Act, 48 Fed. Reg. 4972 (February 3, 1983). The agency subsequently reversed its position and interpreted Part D of the Clean Air Act to require states only to submit plans projecting attainment of the NAAQS by the statutory dates, not necessarily to actually attain the standards by those dates. Compliance with the Statutory Provisions of Part D of the Clean Air Act, 48 Fed. Reg. 50,686, 50,689 (November 2, 1983). EPA took the position that "SIPs were only expected to 'provide for' attainment in a prospective or planning sense" and that approval of a Part D SIP fully satisfied the state's obligations under the 1977 Clean Air Act. Id. at 50,691; see 40 C.F.R. 52. 24(a) (1994) (providing that the construction ban in Section 110(a)(2)(I) does not apply to any nonattainment area once EPA has fully approved the SIP for the area as meeting the requirements of Part D). Thus, the construction ban was "an inducement to timely planning, rather than a penalty for unsuccessful planning." Where an approved plan failed to achieve NAAQS by the end of 1982, however, EPA would treat the plan as "substantially inadequate" and would call for a SIP revision under Section 110(a)(2)(II) (requiring each SIP to contain provisions for revision in the event the EPA issues a SIP call). States that failed to respond to the SIP call would be subject to the construction and funding restrictions in Sections 173(4) and 176(b) for failure to adequately implement their SIPs.

75. R. E. Yuhnke, "The Amendments to Reform Transportation Planning in the Clean Air Act Amendments of 1990," *Tulane Environmental Law Journal* 5 (December 1991): 240-253.

76. State Implementation Plans for Nonattainment Areas for Ozone and Carbon Monoxide, 52 Fed. Reg. 26,404 (July 14, 1987). The EPA initially considered an iterative planning process, allowing states lacking approved SIPs to avoid sanctions by showing reasonable progress toward attainment, but ultimately concluded that this was not consistent with congressional intent, which contemplated an "upfront, complete" planning process. The agency concluded that extension areas that had not submitted a Part D SIP demonstrating attainment by a "near-term, fixed deadline" would be subject to the construction ban in Section 110(a)(2)(I) but that it would not impose any funding sanctions under Section 176(a) as long as the state was making "reasonable efforts" to submit an adequate plan. Likewise, EPA would disapprove any deficient SIP revisions submitted by nonextension areas with approved SIPs in response to the earlier SIP calls and impose the construction ban in Section 173(4) but it would not immediately impose the funding sanctions under Section 176(b). See former 42 U.S.C. §§ 7503(4) & 7506(b).

77. Approval of Post-1987 Ozone and Carbon Monoxide Plan Revisions for Areas Not Attaining the National Ambient Air Quality Standards, 52 Fed. Reg. 45,044 (November 24, 1987). Areas with approved Part D SIPs, such as the Bay Area, would not be subject to the construction ban in Section 110(a)(2)(I) or the funding limitations in Section 176(a). They would, however, be subject to the construction ban in Section 173(4) and funding restrictions under Section 176(b) for failing to implement their approved plans. See 48 Fed. Reg. 50,686, 50,686, 50,693 (November 2, 1983) (sanctions for nonextension areas with approved Part D SIPs that fail to respond to the post-1982 SIP call), discussed in note 74 supra. The DOT took the position that the EPA's authority to impose sanctions expired on December 31, 1987, the last day for demonstrating NAAQS attainment. 53 Fed. Reg. 35,178 (September 9, 1988).

78. 52 Fed. Reg. 45,044, 45,049-52, ¶ I. B. 1&2 (November 24, 1987). EPA intended to impose the construction ban in Section 173(4) on any area that failed to adequately respond to a SIP call by submitting a plan showing attainment within a 3- to 5-year period; however, it also proposed to impose a grant cutoff under Section 176(b) only where it would not be counterproductive to good planning. As the EPA explained, "It makes little sense to withdraw financial support from State workers on whom EPA depends for adequate planning to produce an approvable plan in response to a SIP call." See 52 Fed. Reg. 26,404, 26,409-10, ¶ C (July 14,

1987) (disapproving SIP revisions proposed by nonextension areas in response to earlier SIP calls and imposing the ban under Section 173(4) but not imposing any restrictions under 176(b)), discussed in note 76 supra.

79. 52 Fed. Reg. 45,044, 45,052-3, ¶ I. C. 1 (November 24, 1987). Section 110(c) appeared to require the EPA to promulgate a FIP within 60 days if a state failed to respond to a SIP call. See former 42 U.S.C. § 7410(c)(1). EPA interpreted the time period as not commencing until the sanctions had a reasonable opportunity but failed to induce a state to submit an adequate SIP revision, reasoning that if a FIP had to be imposed within 6 months, "it would rarely be the case that the ban, or any other sanction, would be in place for sufficient time to achieve the congressional purpose of inducing the development of adequate State plans."

80. 52 Fed. Reg. 45,044, 45,066, ¶ IV. B. 1 (November 24, 1987). Exceptions were made for any nonattainment areas with ozone design values less than 16 pphm or CO design values below 17 ppm, which could show attainment in the short term.

81. Areas that could not demonstrate attainment within the 3- to 5-year period would have to show a 9% reduction in ozone for each subsequent 3-year period until attainment to satisfy the "reasonable efforts" requirement and avoid additional discretionary sanctions. Id. at 45,067, ¶ IV. B. 2. a. i.

82. Id. at 45,070-1, ¶ IV. B. 2. a. ii.

83. U.S. General Accounting Office, *Air Pollution: EPA Needs More Data From FHWA on Changes to Highway Projects,* March, 1990, 21. The EPA's comments were in response to the DOT's September 9, 1988, Notice of Proposed Rulemaking, designed to consolidate and simplify its conformity procedures and replace the 1980 joint guidance (53 Fed. Reg 35,178). DOT proposed to retain the basic approach of merely comparing transportation plans and programs with the SIP to ensure that they contributed to reasonable progress in implementing TCMs. The EPA urged the DOT to instead return to the conformity principles of the 1975 joint guidance, which were designed to preclude approval of any plans or projects that would prevent or delay attainment of the NAAQS. See discussion in note 52 supra.

84. C. N. Oren, "The Clean Air Act Amendments of 1990: A Bridge to the Future?" *Environmental Law* 21(1991): 1817.

85. 136 Cong. Rec. S16971 (October 27, 1990).

86. Yuhnke, supra note 75 at 245-246. These include withholding federal highway funds other than for safety or environmental projects and imposing a 2-to-1 offset for pollutants from any new or modified sources requiring a permit. See 42 U.S.C.A. § 7509(b) (West 1995).

87. 42 U.S.C.A. § 7511(b)(2) (for Marginal, Moderate and Serious ozone nonattainment areas); 42 U.S.C.A. § 7512(b)(2) (for Moderate CO nonattainment areas) (West 1995). States that anticipate they will be unable to meet the standards are encouraged to make an early request for reclassification to avoid sanctions or having a federal implementation plan imposed on them. See General Preamble for the Implementation of Title I of the Clean Air Act, 57 Fed. Reg. 13,499, 13506 (April 16, 1992) [hereinafter "General Preamble"].

88. 42 U.S.C.A. § 7502(c)(3) (West 1995).

89. CO nonattainment areas classified as Serious or Moderate with design values above 12.7 ppm were to submit attainment demonstrations, including a control strategy, by November 15, 1992. 42 U.S.C.A. § 7512a(a)(7) (West 1995). Moderate ozone nonattainment areas had to submit their NAAQS attainment demonstrations by November 15, 1993. 42 U.S.C.A. § 7511a(b)(1)(A) (West 1995) (those using regional airshed modeling had until November 15, 1994). Serious and above ozone nonattainment areas had to submit an attainment demonstration by November 15, 1994, using photochemical dispersion modeling or other method approved by the EPA. 42 U.S.C.A. § 7511a(c)(2)(A) (West 1995). See General Preamble, supra note 87, at 13,506.

90. 42 U.S.C.A. § 7502(c)(1) (West 1995). Congressional sponsors apparently intended the transportation provisions of the 1990 amendments to codify the EPA's earlier rule making as approved by the decision in *Delaney v. EPA*. 136 Cong. Rec. S16971 (daily ed. Oct. 27, 1990). See note 50 supra. The EPA believes, however, that position was abandoned in the final bill. On the basis of its experience with implementing TCMs, which led it to realize that local circumstances vary to such a degree that it was inappropriate to presume that all Section 108(f) measures would be reasonably available to all areas, the EPA in 1992 modified its earlier policy of requiring areas to specifically justify a determination that any Section 108(f) measure was not reasonably available. The EPA concluded, "It is more appropriate for States to consider TCM's on an area-specific, not national, basis and to consider groups of interacting measures, rather than individual measures." General Preamble, supra note 87, at 13,561. Under the new regulations, all states must at least address the Section 108(f) measures. The EPA interprets *Delaney* as having been based on its own earlier 1979 policy (rather then the Clean Air Act itself) and thus as not precluding the agency from changing that policy.

91. 42 U.S.C.A. § 7502(c)(9) (West 1995). This means states must show that the measures can be implemented with minimal further action on their part and with no additional rule-making actions, such as public hearings or legislative review. General Preamble, supra note 87, at 13,512.

92. J. A. Gibson, "The Roads Less Traveled?: Motoring and the Clean Air Act Amendments of 1990," *Natural Resources & Environment* 7 (Fall 1992): 15.

93. Section 187(a)(2)(A) provides,

> No later than 2 years after November 15, 1990, for areas with a design value above 12.7 ppm at the time of classification, the plan revision shall contain a forecast of vehicle miles traveled in the nonattainment area concerned for each year before the year in which the plan projects the national ambient air quality standard for carbon monoxide to be attained in the area. The forecast shall be based on guidance which shall be published by the Administrator, in consultation with the Secretary of Transportation, within 6 months after November 15, 1990. The plan revision shall provide for annual updates of the forecasts to be submitted to the Administrator together with annual reports regarding the extent to which such forecasts proved to be accurate. Such annual reports shall contain estimates of actual vehicle miles traveled in each year for which a forecast was required.
>
> 42 U.S.C.A. § 7512a(a)(2)(A) (West 1995).

94. 42 U.S.C.A. § 7512a(a)(3) (West 1995).

95. Section 182(d)(1)(A) states in part,

> Within 2 years after November 15, 1990, the State shall submit a revision that identifies and adopts specific enforceable transportation control strategies and transportation control measures to offset any growth in emissions from growth in vehicle miles traveled or numbers of vehicle trips in such areas and to attain reduction in motor vehicle emissions as necessary, in combination with other emission reduction requirements of this subpart, to comply with the requirements of subsection (b)(2)(B) and (c)(2)(B) of this section (pertaining to periodic emissions reduction requirements).
>
> 42 U.S.C.A. § 7511a(d)(1)(A) (West 1995). § Section 187(b)(2) provides in part,
>
> Within 2 years after November 15, 1990 the State shall submit a revision that includes the transportation control measures as required in [Section 182(d)(1)] except that

such revision shall be for the purpose of reducing CO emissions rather than volatile organic compound emissions.

42 U.S.C.A. § 7512a(b)(2) (West 1995). See Gibson, supra note 92, at 15.

96. H. W. Waxman, G. S. Wetstone, and P. S. Barnett, "Roadmap to Title I of the Clean Air Act Amendments of 1990: Bringing Blue Skies Back to America's Cities," *Environmental Law* 21 (1991): 1873.

97. Yuhnke, supra note 75, at 247. Between 1983 and 1990, VMT increased by 41% nationwide. U. S. Department of Transportation, *Clean Air Through Transportation: Challenges in Meeting National Air Quality Standards* (August 1993), 5.

98. For a discussion of the land use/transportation/air quality ("LUTRAQ") connection between the 1990 amendments and ISTEA, see Netter and Wickersham, "Driving to Extremes: Planning to Minimize the Air Pollution Impacts of Cars and Trucks (Parts I & II)," *Zoning and Planning Law Report* 16 (September & October 1993): 145-151, 153-159.

99. 23 U.S.C.A. § 134(g) & (h) (West Supp. 1995).

100. 23 U.S.C.A. § 134(g)(2)(A) (West Supp. 1995).

101. 23 U.S.C.A. § 134(h)(2)(A) (West Supp. 1995).

102. 23 U.S.C.A. § 134(i) (West Supp. 1995).

103. 23 U.S.C.A. § 134(b) (West Supp. 1995). This requirement applies only to MPOs re-designated after the date of enactment of ISTEA.

104. Conformity Rule, supra note 59, at 62,214.

105. 23 U.S.C.A. § 134(l) (West Supp. 1995).

106. 23 U.S.C.A. § 149 (West Supp. 1995). See generally, U.S. Department of Transportation, *Intermodal Surface Transportation Efficiency Act of 1991: A Summary,* not dated.

107. 23 U.S.C.A. § 134(g)(3) (West Supp. 1995); 42 U.S.C.A. § 7504(b) (West 1995).

108. 42 U.S.C.A. § 7506(c)(2)(A) (West 1995).

109. 42 U.S.C.A. § 7506(c)(2)(B) (West 1995); see also Conformity Rule, supra note 59, at 62,197-8. Only those TCMs that are eligible for federal funding must meet the "timely implementation" test. Id. at 62,211.

110. 42 U.S.C.A. § 7506(c) (West 1995). In turn, the SIP must be coordinated with the RTP. 42 U.S.C.A. § 7504(b) (West Supp. 1995); see also 23 U.S.C.A. § 134(g)(3) (West Supp. 1995) (RTP must be coordinated with the TCMs in the SIP).

111. 42 U.S.C.A. § 7506(c)(1)(B) (West 1995).

112. 42 U.S.C.A. § 7506(c)(2)(A) (West 1995).

113. Yuhnke, supra note 75, at 247.

114. Conformity Rule, supra note 59, at 62,190.

115. U.S. Department of Transportation, *Clean Air Through Transportation: Challenges in Meeting National Air Quality Standards* (August 1993), App. F.

116. Id. at 34-39.

117. Id. at 15.

118. Gibson, supra note 92, at 51. A recent study of a sample of eight nonattainment areas that submitted SIP revisions to the EPA found, however, that TCMs accounted for only a small portion of the projected emission reductions, citing concerns over cost-effectiveness and political acceptability. J. P. Anderson, and A. M. Howett, *SIPs, Sanctions and Conformity: The Clean Air Act and Transportation Planning* (Harvard University, John F. Kennedy School of Government, Taubman Center for State and Local Government, June 1995).

The Bay Area Lawsuit

In June 1989, two environmental organizations filed separate actions in San Francisco federal district court, claiming that the State of California and various regional planning agencies had violated the provisions of the 1977 Clean Air Act by not doing all they could to meet the clean air standards. The first suit, titled *Citizens for a Better Environment v. Deukmejian, et al.,* was brought by a statewide nonprofit environmental organization, headquartered in San Francisco, with more than 25,000 members interested in the protection and improvement of air quality. Their complaint primarily concentrated on the region's failure to curb stationary sources of pollution. The second suit was filed by the Sierra Club, a well-known national conservation organization, with more than 495,000 members, many of whom live or work in the Bay Area and other areas affected by air pollution in the San Francisco Basin. This action, *Sierra Club v. Metropolitan Transportation Commission, et al.,* focused more on the transportation side of the issue. The Sierra Club had previously filed similar lawsuits over air quality in Los Angeles and Sacramento, in one case obtaining a federal court order directing the Environmental Protection Agency (EPA) to tighten regulations on automobile pollution in Southern California.

Together, Citizens for a Better Environment (CBE) and the Sierra Club sued the Bay Area Air Quality Management District (District), the Metropolitan Transportation Commission (MTC), and the California Air Resources Board (ARB), a state agency. The CBE lawsuit also named as defendants the Association of Bay Area Governments (ABAG) and the governor of California. Although both lawsuits addressed the purported failures of the state and local agencies to improve air quality in the San Francisco Bay Area, the Sierra Club's action also charged that the federal EPA had not adequately performed its oversight role to see that the other defendants met their legal responsibilities under the Clear Air Act and the provisions of the approved State Implementation Plan (SIP). As noted earlier, both the EPA and Congress were well aware of the failure of the Bay Area and other regions to attain the National Ambient Air Quality Standards (NAAQS) and were taking steps at the time to strengthen enforcement of the Clean Air Act. The EPA had already informed the Bay Area that it was in violation and had ordered it to submit a SIP revision. Although the Sierra Club obviously intended the lawsuit to put more pressure on the federal government, its objectives were similar and the claims against the EPA were eventually dismissed.

Both actions were assigned to Judge Thelton E. Henderson, who later consolidated them into a single case based on the similarity of the issues raised and the overlap in named defendants.[1] Judge Henderson received his law degree in 1962 from the University of California and served briefly as a U.S. Attorney in the Department of Justice before entering private practice in San Francisco. President Carter appointed him to the federal bench in 1980. At the time this case was filed, he was handling both civil and criminal cases and had earned a reputation for careful deliberation and evenhandedness. He had already ruled on several high-profile discrimination cases and was known for his courteous but firm courtroom demeanor and for giving both sides ample opportunity to present their arguments and explain their legal positions. During the litigation, Judge Henderson was appointed Chief Judge of the Northern District, the first African American to hold that position.

The basis for the plaintiffs' lawsuits was that the various defendants all had some degree of responsibility for developing, approving, and implementing the SIP for the Bay Area but had failed to achieve the reductions in air pollution called for in that plan and had failed to carry out specific portions of the plan designed to improve air quality. Plaintiffs argued that the court should require the planning agencies to implement all the components of their regional air quality plan. The defendants argued that the plan's provisions were advisory and should be treated with flexibility and should not be literally or rigidly interpreted. The district court, although flexible and interpretive in some ways, held that the plan constituted a set of relatively firm commitments that the state and local defendants would have to follow.

Specifically, the court ruled that the defendants had to implement all the control measures listed in the SIP and also had to identify and adopt additional contingency measures when it became clear that the Bay Area would not achieve the federal air quality standards. In addition, the court decided that the MTC's methods for ensuring that highway projects in its regional transportation plans and programs conformed to the SIP were insufficient, and ordered the MTC to delay work on some existing projects and postpone decisions on any significant new projects until adequate procedures were developed. These rulings have already caused planners all over the country to be much more circumspect about the provisions they include in their transportation plans. Before turning to the specific legal issues raised in the case, it will be useful to review the background leading up to the adoption of the air quality plan that was under attack and to describe some of its key elements.

The Bay Area Air Quality Plan

California failed to file an implementation plan for the San Francisco Bay Area by the January 1, 1979, deadline. The nine-county Bay Area Basin[2] was, at the time, a nonattainment area for both ozone and carbon monoxide (CO) although its pollution problems were far less serious than those in other portions of the state, such as Southern California.[3] Sanctions, including a moratorium on the construction of new major sources of ozone or CO in the basin, were imposed by the EPA.[4] Due to difficulties in meeting the federal timetable, the state requested an extension beyond 1982, as permitted under the 1977 amendments, for attaining both the ozone and CO standards.[5] As noted in Chapter 1, the state was obliged to submit an implementation plan revision by July 1979 to qualify for the extension. The 1979 San Francisco Bay Area Air Basin Nonattainment Plan, which was subsequently adopted as part of the state SIP, was designed to attain the federal ozone standards by 1985. The plan relied on four major elements:

1. Use of available pollution control technology on existing stationary sources
2. New source review
3. Motor vehicle inspection and maintenance (I/M) program
4. Transportation system improvements.

The EPA, however, never approved the 1979 Nonattainment Plan because the agency determined that the California state legislature had not authorized the I/M program. The plan was formally disapproved in March 1982.[6]

In 1980, the ARB, which was responsible for adopting the SIP and submitting it to the EPA,[7] designated the three local defendants—the District, the MTC, and ABAG—to prepare the next set of revisions, which were due by July 1, 1982.[8] The District is a local agency set up under state law and is responsible for the adoption and enforcement of rules to achieve and maintain air quality standards in the Bay Area.[9] As noted earlier, the MTC is a regional transportation planning agency created by the California legislature to provide comprehensive transportation planning for the Bay Area, including public and mass transit, highways and roads. It is also the designated regional metropolitan planning organization (MPO) and is responsible for analyzing the air quality impacts of federally funded highway projects for conformity with the local SIP and the Clean Air Act.[10] The third defendant, ABAG, is the regional planning agency for the San Francisco area.[11] It is one of a number of Councils of Governments (COGs) in the state and was established under a joint powers agreement between the various cities and counties in the Bay Area. ABAG is responsible for overseeing the San Francisco Environmental Management Plan[12] and for preparing population and housing growth projections for the region to be used by other local planning authorities.[13]

Pollution Control Measures

The revised plan was finalized in October 1982. As approved by the EPA in January 1984, the 1982 Bay Area Air Quality Plan (Bay Area Plan or the Plan) provided for attainment of CO and ozone standards by 1987 through a combination of stationary source and transportation-related control measures.[14] It also provided for maintaining reasonable further progress (RFP) toward attainment in the interim period. Because the state had subsequently adopted an I/M program, the EPA removed the sanctions it had imposed on new construction.[15]

Specifically, the 1982 Bay Area Plan provided for a total of 23 new stationary source controls designed to reduce hydrocarbon (HC) emissions and thereby lower ozone levels in the region. Eleven additional measures were placed in the contingency category, and a process was established to reevaluate these measures in the event further reductions from stationary sources became necessary. A key component of the Plan continued to be the statewide I/M program, which finally went into effect that September. This biennial smog inspection program was expected to achieve a 25% reduction in vehicle emissions.

Land use control measures were considered only marginally. The Plan specifically rejected regional land use controls; instead, it merely recommended that cities and counties should consider modifying their general plans to "contain development in urban service areas with urban services in place; encourage mixed-use development and infill on vacant land at densities sufficient to support transit; and encourage rehabilitation and reuse of older buildings."[16]

The Bay Area Plan also established specific targets for reducing allowable pollution levels in the Basin. As described in Chapter 1, the NAAQS are expressed in terms of the allowable atmospheric concentration of each pollutant, so many parts per million (ppm) or parts per hundred million (pphm). For purposes of pollution control strategies, it is necessary to convert these figures into estimated emissions levels, usually given as so many tons per day. As long as the daily emissions do not exceed these levels, it can be assumed that the pollutant concentration in the air will stay below the federal limits. The Plan contained separate recommendations for reducing daily emissions levels of both ozone and CO to meet the standards. These became a major focus of the lawsuit, with the plaintiffs arguing that the defendants were legally obligated to achieve the reductions specified in the Plan, so it is important to understand exactly what the Plan said about them.

Ozone

As noted earlier, ozone is produced through a chemical reaction involving certain hydrocarbons (HCs) and oxides of nitrogen (NOx). The Bay Area

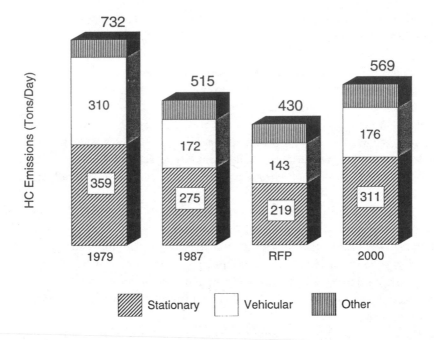

Figure 2.1. Baseline Hydrocarbon Emissions

established a baseline for HC emissions for the years 1979, 1987, and 2000.[17] As shown in Figure 2.1, estimated HC emissions from all sources for 1979, represented by the far left column, were 732 tons per day (tpd). Stationary sources of pollution, such as factories, contributed 359 tpd to the total. Vehicle emissions accounted for another 310 tpd. The remainder came from other sources, such as nonroad vehicles like boats and lawn mowers. The middle left column represents projected HC pollution levels in 1987. Emissions levels were expected to decline to 515 tpd by that time, assuming no further actions were taken beyond the existing pollution control measures already in place. For convenience, these are referred to as *uncontrolled* emissions. After that, pollution levels were expected to rise as the effects of regional population growth would begin to overwhelm existing controls. By 2000, assuming no further controls, a total of 569 tpd in HC emissions was projected, as indicated by the far right column, and the share attributed solely to vehicles was expected to increase slightly from 172 tpd to 176 tpd.

The Bay Area Plan called for a combination of new control measures to reduce total HC emissions for the Bay Area Basin another 85 tons by 1987. This reduction was deemed necessary to reduce the expected ozone concentrations down to the federal standard. MTC used the Livermore Regional Air Quality (LIRAQ) model to simulate the ambient ozone concentration or *design value* associated with these baseline emissions.[18] The 1979 emissions were calculated to be equivalent to a design value of 19 pphm. For 1987, the design value was 14.4 pphm. Again, the NAAQS for ozone is 12 pphm. Because the 1987 design value exceeded the federal standard by about 17% and was associated with an estimated 515 tpd emissions level, this meant daily emissions would have to be reduced by a similar percentage to attain NAAQS for ozone ($515 \times 0.17 = 85$).[19] Thus, as illustrated by the middle right column in the figure, RFP would be demonstrated by gradually reducing all HC emissions down to 430 tpd (515 minus 85) by 1987.[20] According to the Plan, the stationary source sector would be responsible for reducing vehicle emissions by 56 of the 85 tons. The transportation sector was responsible for achieving the remaining 29 tons, chiefly through the I/M program.[21]

Carbon Monoxide

The Bay Area Plan also contained strategies for reducing CO pollution. The major problem area for CO is in the Santa Clara Valley, particularly downtown San Jose, which is located south of the San Francisco Bay. The Plan included estimates for regional CO emissions. Total estimated emissions from all sources for 1979 were 3,220 tpd, as illustrated in Figure 2.2. Projections showed that with existing controls in place, regional CO emissions would decline to 2,340 tpd by 1987. In contrast to HC emissions, which were anticipated to increase after 1987, CO emissions would continue to fall slightly to 2,250 tpd by the year 2000, even without additional measures.

The Plan relied solely on the transportation sector to achieve additional CO reductions. The statewide I/M program, administered by the ARB, was expected to reduce emissions by 367 tpd.[22] Ten transportation control measures (the 10 TCMs) were also proposed to further reduce CO emissions in the region.[23] Table 2.1 lists these 10 TCMs and their expected contributions to NOx, HC, and CO emissions reductions, although no credit was claimed for them toward ozone RFP in the Plan. The MTC concluded that any

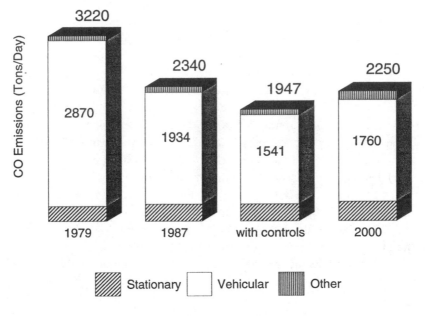

Figure 2.2. Baseline Regional Carbon Monoxide Emissions

additional Section 108(f) TCMs would not be effective, and none were included in the Plan.[24]

On top of the reductions from the I/M program, the 26 tons from the 10 TCMs would produce a total of 393 tpd in pollution savings. Because the *uncontrolled* CO emissions from the transportation sector alone would be 1,934 tpd by 1987, as shown by the middle left column in Figure 2.2, the planned reductions would leave a projected residual of 1,541 tpd (1,934 minus 393), as shown by the middle right column.

Because CO pollution is primarily a local rather than a regional phenomenon, the Bay Area Plan focused on the three smaller areas with the highest potential for developing CO hot spots—the cities of San Jose, Oakland, and Vallejo. Again, the projections showed that CO levels were expected to decrease by 1987, due mainly to existing motor vehicle control programs. Attainment was projected for Oakland and Vallejo, although for downtown San Jose the decrease would not be sufficient to counter increased emissions due to higher traffic volumes, and as a result the federal 8-hour standard would be exceeded in that area.

TABLE 2.1 Emissions Reductions From 10 TCMs

Transportation Control Measure	*Reductions (Tons/Day)*		
	NOx	*HC*	*CO*
1. Commitment to 28% increase in transit ridership 1978-1983	—	—	—
2. Support improvements in transit operators 5-year plans to increase ridership	1.04	0.72	7.15
3. Seek to expand transit beyond committed levels	0.54	0.37	3.69
4. Support development of HOV[a] lanes	—	—	—
5. Support RIDES effort[b]	—	—	—
6. Continue efforts to support long-range transit improvements	—	—	—
7. Reaffirm commitment to preferential parking programs	—	—	—
8. Park-and-ride lots	0.05	0.04	0.19
9. Expand commute alternatives program	0.89	0.87	8.83
10. Develop information program for local governments	0.27	0.69	6.04
Total	2.79	2.69	25.9

a. High-occupancy vehicle.
b. RIDES is a nonprofit corporation funded by MTC and the California Department of Transportation that provides car pool matching services in the Bay Area and also facilitates vanpooling.

Because there were high background levels of CO in the urbanized areas of San Jose, a program of simple traffic redistribution to reduce local road congestion appeared unlikely to produce significant air quality benefits. Therefore, the Plan called for steps to be taken to reduce vehicle emissions on a wider subregional scale. Twelve transportation mitigation measures were considered for the San Jose urban area. The Plan recommended adopting a Commute Transportation Program and a Gasoline Conservation Aware-

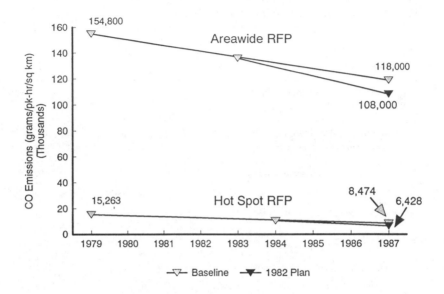

Figure 2.3. Subregional CO Emissions, 1979 to 1987

ness Program while listing a light rail transit system as a contingency measure.[25] These efforts were expected to eliminate any potential hot spots.

The RFP schedule for CO in the Bay Area Plan established standards for *Areawide* CO emissions in a 25 sq. km. region around San Jose and for a sample *Hot Spot* location at the intersection of First and Santa Clara streets in downtown San Jose.[26] A 28% decrease in CO emissions was needed at this design intersection in order for all intersections in the downtown San Jose study area to meet federal standards.[27] The shaded upper line in Figure 2.3 represents Areawide uncontrolled emissions, which were projected to go from 154,800 grams per peak hour per square kilometer (gm/pk-hr/km^2) in 1979 down to 118,000 gm/pk-hr/km^2 by 1987. The I/M Program and the Commute Transportation Program were expected to further reduce emissions. Similarly, Hot Spot emissions represented by the lower shaded line in the figure, were expected to drop from 15,263 gm/pk-hr to 8,474 gm/pk-hr. Here, additional credit was taken only for reductions from the I/M Program.

With the all the control measures in place, Areawide CO emissions would be reduced to 108,000 gm/pk-hr/km^2 and Hot Spot emissions to 6,428 gm/pk-hr, as illustrated by the two solid black lines in Figure 2.3. A program

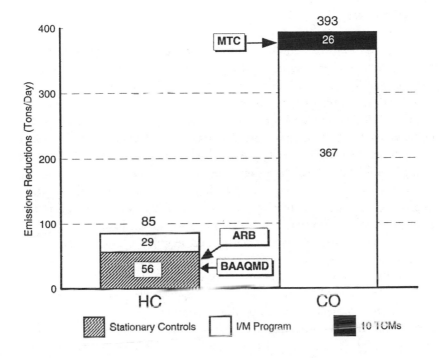

Figure 2.4. Bay Area Air Quality Plan Pollution Control Programs

to monitor RFP would also be developed, including traffic counts at key intersections in downtown San Jose and an assessment of the effectiveness of the Commute Transportation Program.[28]

To summarize, the basic approach of the Bay Area Plan was to use any reasonably available methods to control the three principal sources of air pollution: stationary sources, mobile sources, and transportation-related sources. The stationary source sector was responsible for lowering ozone emissions, but the transportation sector had to reduce both HC and CO. The Plan contained 23 stationary source control measures for ozone, a statewide mobile source control program setting new car motor vehicle emission control standards that addressed both ozone and CO emissions and required a biannual smog inspection, and regional TCMs to reduce CO emissions. As illustrated in Figure 2.4, the ARB and the District were jointly responsible for controlling stationary ozone emissions, and the MTC was responsible for all transportation system-related measures.

Conformity Review

In addition to these pollution control measures, the Bay Area Plan contained provisions for ensuring conformity between transportation planning and air quality goals to comply with the provisions of Section 176(c) prohibiting any MPO from approving any nonconforming "project, program, or plan." MTC, as the local MPO, is responsible for preparing and annually updating the regional transportation plan (RTP) for all existing and proposed transportation systems in the Bay Area.[29] As previously explained, the RTP is a *planning* document that establishes both long-range and short-range goals and describes programs, including highway construction, necessary to implement the plan, together with a financial analysis covering projected costs and available revenues.

MTC is also responsible for preparing and adopting a Regional Transportation Improvement Program (RTIP), which contains transportation projects that local government sponsors wish to have included in the State Transportation Improvement Program (STIP), the statewide program of highway and transportation projects prepared by the California Transportation Commission. The STIP sets forth the projects the state intends to construct with state or federal highway funds. Projects approved for the STIP are then eligible for inclusion in the federally required transportation improvement program (TIP) for the Bay Area. Again, the TIP is a *programming* document that covers a 3-year period and includes all projects eligible for funding and implementation during that period. Projects included in the TIP are not necessarily constructed within the covered 3-year period but may be delayed due to funding, environmental, or other concerns.

Appendix H to the Bay Area Plan contained provisions, in accordance with the EPA's 1981 SIP approval policy, to ensure "that transportation plans, programs and projects approved by [the MTC] are in conformance with the SIP."[30] Specifically, it provided that all RTP amendments and the biannual TIPs would be reviewed to determine their air quality impacts and the effect of all adopted TCMs. Appendix H also specifically required the MTC to assess all "major highway projects" in the TIP to see if they might adversely affect emissions. The Plan did not, however, specify how these assessments should be made. The full text of the Conformity Assessment is reproduced in Appendix A.

Despite adoption of the Bay Area Plan, the region had still not achieved the NAAQS by the extended deadline. Several of the 23 proposed stationary source control measures, those governing large bakeries, auto finishing, consumer solvents, pesticides, and small gasoline engines, and together accounting for about 20% of the projected reductions, had never been adopted. In May 1988, the EPA notified the state that the Bay Area Plan was inadequate to attain the NAAQS, and in June ordered it to submit schedules to adopt all measures committed to in the Plan but not yet enacted. The ARB, however, chose to postpone hearings on the bakery rule, and the District decided to delay action on an auto finishing rule pending further hearings. By February 1989, though, no action on adopting the remaining measures had been taken either.[31] The EPA also took steps to enforce the contingency plan.

The Contingency Plan

As noted in Chapter 1, EPA regulations implementing the 1977 Clean Air Act required SIPs to contain contingency measures in the event that sufficient progress toward attaining the NAAQS could not be achieved.[32] Accordingly, the Bay Area Plan contained a Contingency Plan that called for the implementation of additional measures to further reduce emissions from both the transportation and the stationary source sectors. The stationary source portion included consideration of the 11 contingency controls mentioned earlier.[33] The transportation portion of the Contingency Plan called for the MTC to

1. consider delaying any highway projects shown to have a "significant adverse impact" on air quality and
2. adopt additional TCMs within 6 months of a determination that RFP was not being met to bring the region back into compliance.

Annual RFP reports were to be submitted each July to monitor the region's performance in implementing the TCMs.[34] The text of the Transportation Contingency Plan is reproduced in Appendix B.

The RFP reports were not submitted on time, as required, and those that were later produced indicated that the region had fallen behind the schedule set out in the Plan for reducing HC and CO emissions. Moreover, the emissions reductions from the I/M program were not nearly as large as initially projected. By February 1989, although the District had adopted several new contingency measures for ozone, the MTC had not even considered any measures to reduce CO emissions. Because the purpose of the Contingency Plan was to correct shortfalls in emission reductions from the transportation sector to bring the region "back within the RFP line," the EPA ordered the agency to "evaluate, adopt and implement" additional TCMs as soon as possible to fulfill its legal obligations. EPA also threatened sanctions for failing to implement the Plan, including a construction ban and the loss of federal funds, if the state did not respond.[35]

At the same time, CBE and the Sierra Club prepared to challenge the government's failure to implement contingency measures to meet the clean air standards. In accordance with federal law, the environmental organizations served the defendants with notices of intent to sue, specifying the matters on which they sought action. They asserted that the District and the ARB had failed to adopt all the stationary controls listed in the Plan and had failed to identify and adopt stationary source contingency measures to attain federal ozone standards by the 1987 deadline. They also charged that the MTC had neglected to carry out the transportation portion of the Contingency Plan by not delaying highway projects and not adopting additional TCMs. In addition, CBE and the Sierra Club claimed that the MTC had failed to conduct an analysis of the air quality impacts of any RTP amendments and/or an assessment of whether any planned highway projects in the TIP would adversely affect emissions as required by the Transportation Conformity Assessment. Finally, the Sierra Club also accused the EPA of failing to enforce the SIP by not imposing a moratorium on new permits for stationary sources of air pollution, not establishing a set date for a SIP revision once the clean air standards were not met, and failing to prepare a Federal Implementation Plan (FIP) for the Bay Area.

When no action was taken by the government agencies within the 60-day response period provided by law,[36] the environmental organizations filed the aforementioned suits to bring the area into compliance and enforce the provisions of the Bay Area Plan, only days after President Bush had announced his proposals for a new Clean Air Act. Initially, the defendant agencies did not appear to take the lawsuits very seriously. Noting that the Bay Area was a leader in clean air standards, Larry Dahms, executive director of the MTC, insisted that the defendants were already doing most of what the plaintiffs were seeking through their complaints. And although the plaintiffs suggested they would press the court to impose sanctions against the defendants, including halting some highway construction projects, agency officials publicly doubted that the court would ever go along.[37]

The plaintiffs moved quickly for summary judgment against the state and local defendants on their claims, a streamlined procedure for deciding pure questions of law where the basic underlying facts are not in dispute. Lawyers for the various defendants responded by filing opposition briefs, and a hearing date was set in September 1989. The Bay Area lawsuit would prove to be an important test of both the 1977 act and what were to become the 1990 amendments, which took effect during the course of the litigation. The first question to be addressed by the court was whether the defendants could be held legally accountable for the region's failure to achieve the NAAQS. Although, generally speaking, the court held that they could, determining the scope of that liability was not a straightforward matter. The plaintiffs' success depended on a careful reading of the provisions of the Clean Air Act. That issue is taken up next.

Court Ruling on Liability Issues

Broadly speaking, the dispute between the environmental plaintiffs and the government defendants in this case centered on a fundamental difference of opinion over the nature and the purpose of planning. Throughout the 1950s and 1960s, the typical view of plans was that they served as a general set of guidelines to help decision makers formulate policy. For the most part, they were used internally and were not always widely available to the public. In fact, the public typically had little or no role in their preparation, because planning was considered a job for experts. Plans were not considered legally binding documents—that is, they did not obligate officials to undertake any particular actions—instead, they were merely advisory in nature. Basically, they identified long-range *goals* and *objectives* and described various options that might be considered to achieve them. Insofar as they espoused *policies,* the plans tended to be bland or uncontroversial, incorporating such statements as "support highway development" or "expand public transit." Plans were perceived as *static* documents that were intended to operate over a long period of time and be revised only when they became clearly outdated. As a result, many plans ended up on the shelf, collecting dust and rarely having any direct impact on public policy.

In recent years, this view of planning has been changing. Now, plans are more and more expected to operate in the short term and to provide specific guidance for solving particular problems. They are seen as a blueprint for getting from Point A to Point B and representing a commitment on the part of responsible government officials to carry out identified tasks to achieve certain results. They are often the products of considerable public participation and comment and may be widely distributed to the community at large. In a sense, they have come to be thought of not just as recommendations but rather as concrete programs for action and, in some cases, even a sort of "contract" between various concerned groups and the government. Today's plans often contain more specific policies and detailed implementation programs than in the past. They also frequently include quantified standards for measuring performance and requirements for monitoring whether the stated goals and objectives are being achieved within the expected time frames. In addition, plans may provide for periodic review and revision and often contain contingency provisions in the event that the specified benchmarks are not being met. In short, plans are expected to be more dynamic and able to respond to changing conditions, and more binding on those who prepared them.

In this case, the MTC and its codefendants tended to take the older, more institutional view of planning, arguing that the Bay Area Plan was never intended to constrain their authority to undertake highway projects to meet the needs of a growing population. The environmental plaintiffs, on the other hand, considered the Plan to have more "teeth" and to require the defendants to accomplish specific goals. The first phase of the litigation centered on which of these two competing views best reflected the intent of the Clean Air Act. In other words, whether under the law the Bay Area Plan should be viewed as a general set of advisory statements or as a series of commitments to achieve specific clean air objectives.

In their combined summary judgment motions, the plaintiffs alleged that the defendants had failed to meet the Clean Air Act standards and had failed to take corrective actions following that failure as set out in the Plan.[38] Specifically, plaintiffs challenged the District's and the ARB's failure to adopt all 23 stationary source controls identified in the Plan and to adopt contingency measures covering stationary sources of pollution. Plaintiffs also charged the MTC with (a) failure to implement the Contingency Plan after RFP had not been met, and (b) failure to properly determine whether the annual RTP amendments and the TIPs conformed with the Bay Area Plan.

In the first of what were to be several written opinions over the course of the litigation, on March 5, 1990, Judge Henderson granted in part the plaintiffs' motions. He held that once the EPA approved the Bay Area Plan, the state was required to carry it out.[39]

As to the stationary source measures, the judge agreed with the plaintiffs that the Bay Area Plan specifically obligated the defendants to adopt all the listed control measures, brushing aside defense arguments that these provisions did not represent real commitments by the District. The court ruled that making "reasonable efforts" to achieve compliance was insufficient—the legal test was whether the defendant agencies had "in fact" complied with the Bay Area Plan.[40] Because no control measures had ever been adopted for 4 of the 23 listed stationary sources of pollution, the court found that the liability of both the state and local agency defendants was "beyond question." It ordered the District and the ARB to forthwith implement the four missing control measures.[41] The court then turned to the question of whether the defendants had become obligated to carry out the Contingency Plan provisions.

Failure to Adopt Contingency Plan

Finding the defendants liable for not implementing the Contingency Plan proved more complex. Unlike the four missing stationary source controls, which had been clearly spelled out in the Plan, the defendants' obligations to adopt additional contingency measures were not quite as explicit. The major point of dispute was whether the plaintiffs could force the defendants to adopt new regulations to reduce pollution levels beyond those already listed in the Plan.

Consistent with the emerging view of planning described earlier, the plaintiffs viewed the Contingency Plan as an important part of the defendants' commitment to improving air quality. They felt its role was to identify additional steps to be taken in the event the programs and policies in the Bay Area Plan did not achieve the expected emissions reductions. In contrast, the defendants saw it more as a technical requirement that had to be met—for them it merely identified some possible directions that could be taken in the event the other measures in the Plan could not be implemented as planned or proved to be less effective than expected. In their view, it did not obligate them to take any specific actions. Given the ambiguity in the function of the

Contingency Plan, the legal issue turned on whether federal law authorized a suit under these circumstances. The threshold question here for the court was whether the plaintiffs had standing to bring their actions in federal court under the provisions of the Clean Air Act.

Standing to Sue Under the Clean Air Act

The Clean Air Act permits private citizens to bring enforcement actions against governmental agencies. Any person may challenge violations of the act through a "citizen suit" in federal court.[42] However, the law limits the scope of issues that may be litigated. A citizen suit cannot be brought just because a state has failed to attain the NAAQS or because it failed to advance the general policies or other goals contained in a SIP.[43] Nor may such a lawsuit be filed simply over a state's failure to adopt an adequate SIP.[44] The action is available only to enforce an "emission standard or limitation" that is set forth in an approved SIP.[45] The specific provisions subject to enforcement include schedules for compliance, TCMs, vehicle inspection and maintenance plans, air quality maintenance plans, and other controls designed to reduce emissions.[46]

In part, this rule is to prevent outside parties from challenging in court issues that should have been dealt with in the plan approval process. Congress wanted to leave it up to the local authorities to decide the best ways to reduce air pollution. The procedure for preparing and adopting a SIP includes provisions for public comment and for challenging the adequacy of the plan before the EPA. Any legal challenges to the adequacy of the plan must be brought within 60 days of approval by the agency.[47] After that, it is too late to question the workability of the plan. Once the state has identified a particular course of action for achieving NAAQS, it must be given the opportunity to implement its program. However, a state can be held responsible for carrying out those specific steps, the *emissions standards and limitations,* that it determined should be undertaken to achieve cleaner air.

In other words, a court may not require a state to achieve the clean air standards or even to make changes to its existing SIP. It may, however, require a state to follow through on specific commitments already contained in its SIP.[48] Because each state's SIP contains a different set of policies and programs, the issue of what a state can be held responsible for accomplishing therefore depends on the actual language in the SIP. Here, the plaintiffs maintained that the contingency provisions of the Bay Area Plan were in fact such an emissions standard or limitation and that they committed the defendants to taking further steps to attain the NAAQS. The issue for the court to consider, therefore, was whether the Contingency Plan in fact obligated the defendants to take specific actions, and if so, just what actions they required the defendants to take.

Defendants objected that they could not be compelled to adopt new control measures to achieve the NAAQS. They pointed out that although SIPs are designed with the ultimate goal of achieving the federal standards, a SIP does not by itself commit state or local authorities to attaining them.[19] The Bay Area Plan was just a planning tool designed to help achieve that goal. It was, in the agencies' view, within their administrative discretion to decide whether any additional steps should be taken to reduce pollution. The time to challenge the adequacy of the control measures was before the EPA gave its approval to the Plan. Plaintiffs countered that whereas in general a clean air plan might not obligate its authors to achieve NAAQS, here the language of the Contingency Plan specifically committed the defendants to take additional actions to achieve the air quality standards.

Thus, the task for the court here was to decide whether the plaintiffs had identified specific emissions standards and limitations, which the defendants had violated, or whether the plaintiffs were merely trying to force the government to meet general goals contained in the Plan. If the language in the Contingency Plan merely reiterated the general objective to improve air quality, then it would not be *enforceable* because there would be nothing specific for the court to order the defendants to do. In other words, the plaintiffs could not sue just to force the government defendants to undertake unspecified steps to achieve the federal standards for air quality, or to devise new strategies to meet that goal, but they could sue to require the defendants to carry out specific provisions in the Plan that would contribute to achieving those standards. That depended in turn on whether the court interpreted the language in the Plan as mandatory or merely hortatory.

The government defendants argued that (unlike the 23 stationary source controls) the provisions of the Contingency Plan were simply not sufficiently specific to be enforced. In the absence of such specific commitments, they asserted, the court could not just order them to take steps to achieve NAAQS. CBE and the Sierra Club maintained, though, that the EPA requires SIPs to contain enforceable measures and that therefore the Contingency Plan was enforceable, otherwise the Bay Area Plan would not have been approved.

Defendants could not have it both ways—a plan sufficiently concrete to pass muster with the EPA but too vague to be enforced by the courts. The plaintiffs insisted that the Contingency Plan independently required the defendants to achieve NAAQS, making their complaint more than simply an attempt to enforce the statutory requirements. To the plaintiffs, the Contingency Plan represented a specific promise to correct any deficiencies that appeared in the projected emission reductions even if that meant adopting new control measures.

Because the defendants believed that under federal law they could not be directly compelled to attain the NAAQS, in their view the lawsuits simply amounted to an improper attack on the adequacy of the 1982 Plan. To them, the plaintiffs' reasoning was just a backdoor attempt to require the agencies, in effect, to submit a new strategy for attaining the NAAQS for CO and ozone, something that this court could not lawfully order. Only the EPA could call for a SIP revision if the Plan proved inadequate, which in fact it had done. As the EPA pointed out to the court, the Sierra Club could not "bootstrap itself into the SIP revision process by arguing that an enforceable commitment of the 1982 Plan itself requires attainment of the NAAQS."[50]

As noted earlier, the EPA had already notified the state that, based on its annual reports, RFP was not being met and it had directed the state to submit a revised SIP by September 30, 1991. Defendants took the position that instead of fixing any problems in the Bay Area Plan by adopting contingency measures, they should be allowed to address the failure to attain NAAQS as part of this forthcoming SIP revision. Any decision on whether to adopt new control measures should be left to that process, which was by then already under way. In short, the defendants believed that any oversights in the Bay Area Plan were a matter for EPA regulation rather than judicial intervention.

The EPA further pointed out that Congress was aware of the fact that attainment dates had passed and was considering legislative amendments to the Clean Air Act that would establish new attainment dates and require states to adopt new SIPs. It argued that meanwhile the court could not order California to adopt new control measures or delay transportation projects to avoid further violations of the NAAQS. These issues, the EPA contended, were reserved for Congress and the ongoing SIP revision process, described in Chapter 1.

In essence, the plaintiffs viewed the Contingency Plan as an integral part of the regional strategy to achieve cleaner air by correcting any deficiencies in implementing the Bay Area Plan, whether due to faulty assumptions about the efficacy of the chosen control measures or to inaccurate estimates of future pollution levels. The defendants, by contrast, saw it as much less important and held that if the Plan had major shortcomings, the proper response would be to prepare a new plan using current data and methodologies. Against this background, the court considered separately the stationary source contingency provisions and those for the transportation sector.

Stationary Source Contingency Measures

As explained earlier, the District and the ARB were primarily responsible for adopting stationary source control measures. In addition to the 23 measures already mentioned, the Bay Area Plan had identified 11 more potential stationary source controls in the Contingency Plan.[51] It also provided that if emission reductions were not adequate to permit attainment by 1987, then "further controls must be adopted and implemented."[52] Defendants maintained that the Plan did not mandate adopting any particular stationary source measures. They argued that to treat these potential measures as specific commitments, instead of merely suggestions for possible action, would impose a "chilling effect" on the planning process, because it would discourage consideration of other new developments in technology in a SIP for fear that a court might someday insist that these untried methods be adopted simply because they had been included on a list of suggestions for further study. The District also contended that at most the Bay Area Plan committed it only to review and develop selected contingency measures, not necessarily the ones mentioned in the Contingency Plan, and that although it would likely adopt some of the measures now listed in the contingency category, the court should allow it to reevaluate these and other possible measures as part of the revised SIP.

The district court, however, by and large sided with the plaintiffs. In its March 5, 1990 Order, the court held that the Bay Area Plan contained an "unequivocal promise" by defendants to adopt and implement additional stationary source control measures. This "basic commitment" was a "specific strategy" for achieving cleaner air and was enforceable even though no particular measures had been specified. Therefore, the defendants were legally obligated to enact further pollution controls once the region failed to achieve RFP. The court agreed with the defendants, though, that the Plan did not explicitly require enactment of sufficient stationary source measures "such that NAAQS will be met." Instead, it required the defendants only to enact *some* contingency measures after a showing that RFP had not been made. Because the Plan did not link the number of such measures to the attainment of NAAQS, or expressly commit the District to adopt sufficient contingency measures to attain NAAQS, the court felt it could not supply that connection on its own.

Moreover, as the defendants had pointed out, the Clean Air Act does not permit bringing a citizen suit merely to enforce the overall "aims and goals of the [SIP] apart from the specific strategies designed to achieve them."[53] This weighed in favor of narrowly construing the defendants' obligations under the Plan. Because the District had already adopted several contingency measures and had adopted a resolution agreeing to consider additional measures, the court initially refused to find a violation of the Contingency Plan with respect to stationary sources of pollution.

On the plaintiffs' motion for reconsideration, though, the court reversed its original ruling and did find that a violation had occurred.[54] Here, the plaintiffs tried a slightly different argument. They asserted that even if the Bay Area Plan did not directly require the District to attain NAAQS, it did require it to adopt sufficient contingency measures to make reasonable further progress with respect to HC emissions. In other words, under the Contingency Plan, the District was obligated to achieve the RFP targets, something "logically once-removed" from attaining NAAQS themselves although, in theory at least, meeting them should result in the region attaining the federal standards.[55] The court now agreed with this approach, holding that an order directing the defendants to take steps to get back "on the RFP line" would not be the same as an order "enforcing NAAQS directly" but would simply require the defendants to carry out a specific strategy for achieving them, namely producing annual targeted reductions in emissions as measured by the benchmarks shown in the Bay Area Plan for reducing ozone levels. As mentioned earlier, the Plan called for reducing ozone levels to 430 tpd by 1987. The defendants insisted, though, that the ozone RFP targets were merely "estimates and goals, not absolute requirements which had to be met."[56] The court ruled otherwise. It held that the District and the ARB had to meet the RFP target, even if meeting it would not actually result in NAAQS attainment. By requiring the defendants to keep this promise, the court believed it was merely enforcing the progress already committed to in the Plan.

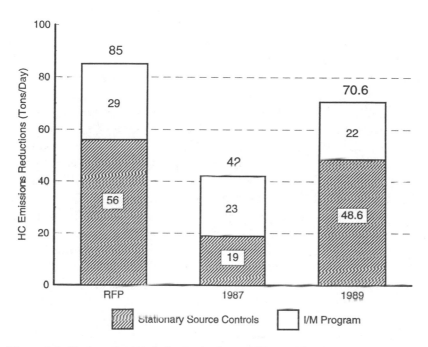

Figure 2.5. Hydrocarbon Emission Reductions, 1987 to 1989

Still, the District and the ARB argued that under the Plan they were responsible only for reducing HC emissions by 85 tpd over what they might otherwise have been, not necessarily for achieving an overall emissions level of 430 tpd by 1987. Indeed, there was some evidence that by 1987 total daily emissions were running as high as 616 tpd due to increased growth in the region.[57] Plaintiffs, on the other hand, insisted that the proper standard was the original 430 tpd figure, especially because the 1987 deadline had already passed. The court concluded that the defendants had failed to achieve RFP, regardless of which standard was used.

The Bay Area Air Quality 1987 RFP Report had stated that only about half the expected HC reductions had been achieved. Overall ozone emissions were 473 tpd or 43 tons short of the 430 tpd standard in the Plan. Breaking this down by stationary and vehicle sources, Figure 2.5 illustrates that emissions reductions from all stationary source controls amounted to only

19 tpd, well short of the 56 tpd target. For the transportation sector, initial estimates suggested that the smog check program achieved just 23 tpd in HC reductions, 6 tons short of projected reductions in the Bay Area Plan.[58] Later estimates showed that even by 1989, emissions reductions from all sources were running only about 70 tpd, still short of achieving the projected 85 ton total.[59]

Given that the District had not adopted all the listed stationary controls and because HC emissions had not been reduced to the RFP target of 430 tpd, nor had the 85 tpd in reductions called for in the Bay Area Plan been achieved (again, the difference between 515 tpd, the level expected based on existing control programs, and the 430 tpd target), the court found the District and the ARB liable for failing to implement their portion of the Contingency Plan.[60] These defendants were found liable for the entire deficit even though some of the overall shortfall in ozone would have to be made up by the transportation sector.[61] The court gave the defendants until December 31, 1991, to carry out their commitment to adopt sufficient additional measures covering stationary sources to demonstrate RFP, measured by *both* the target reductions and the residual emissions level, as they were essentially "two sides of the same coin." Just what this dual requirement meant would be the subject of protracted argument between the parties over the next several years and is discussed in detail in Chapter 4.

Transportation Sector Contingency Measures

Besides mandating additional stationary source controls, the Contingency Plan also provided for emissions reductions in the transportation sector in the event that the region failed to make RFP. As noted earlier, under the Bay Area Plan, the transportation sector was responsible for 29 of the 85 tpd in HC reductions and for the full 393 tpd reduction in regional CO emissions. Beyond the ozone deficiencies described earlier, the belated RFP Reports also showed more than a 200 tpd shortfall in reducing CO emissions.[62] These reports indicated that the deficit was largely due to the failure of the state I/M program to achieve the predicted reductions in ozone and CO emissions.[63] Also, the Bay Area Plan projections for the total number of daily vehicle trips and vehicle miles traveled (VMT) (on which the projections of vehicle emissions were based) had already been exceeded. Both the jobs-housing imbalance in the Bay Area and the strong local economy at that time were cited as factors in the increased vehicle use, along with reduced gasoline prices.[64]

In light of these findings, plaintiffs maintained that MTC was also liable for failing to adopt any contingency measures. Unlike the somewhat vaguer provisions for stationary sources, here the Bay Area Plan affirmatively stated the agency would "adopt additional TCMs within 6 months of the [non-RFP] determination" designed to bring the region back within the RFP line and would "delay certain categories of projects in the TIP" if they were shown to have significant adverse impact on air quality" (see Appendix B). Plaintiffs argued that these two individual elements of the transportation sector part of the Contingency Plan—the contingency TCMs and the procedures for delaying highway construction projects—were specific, enforceable commitments. According to the Sierra Club, MTC had interpreted EPA's failure to enforce sanctions for noncompliance as a license to "continue ignoring requirements and schedules it found impractical or inconvenient" to building more highways.[65] This, they felt, showed the need for citizen enforcement and judicial relief.

Contingency TCMs

Similar to the District's argument over the enforceability of stationary source contingency measures, the MTC argued that the operative language of the Transportation Contingency Plan only set forth a process for deciding what additional transportation control measures might be needed; it did not commit the MTC to adopting any particular measures. Thus, lawyers for the agency argued, the contingency provisions themselves were not an *enforceable TCM,* as that term is used in the Clean Air Act, just because they might lead to the adoption of TCMs in the future. Instead of being specific commitments, they were merely a promise to consider a range of options, if necessary, to maintain further progress toward attainment of the NAAQS. The agency's position was therefore that any such commitments were not sufficiently specific to be enforced, and the question of additional TCMs should again be left to the forthcoming SIP revision. The agency did agree to consider adopting additional TCMs, but only after a review of the public comments, an evaluation of the potential impacts of the proposed TCMs, and a determination of their reasonableness and availability, all required by the Clean Air Act.

The language in the Bay Area Plan, MTC argued, was similar to the commitment addressed in another federal case, *Wilder v. Thomas,* calling for the City of New York to "eliminate CO hot spots" in a section of the city "by the end of 1987." There, the Second Circuit Court of Appeals had held the quoted language would not support a claim against New York City, finding that the provision merely restated the statutory deadline for attaining the federal standard, and therefore the complaint against the City was simply an improper attempt to enforce the NAAQS directly.[66] MTC insisted that the decision in *Wilder* governed this case because the Bay Area Plan merely spoke about taking unspecified steps to achieve RFP; there was no defined strategy to reduce pollution by limiting vehicle use. Though the EPA agreed that the district court had jurisdiction to enforce specific provisions the state had committed to undertake in its SIP (those that constituted emissions standards and limitations), the agency also felt that plaintiffs misinterpreted the SIP requirements in the Clean Air Act in asking the court to order the MTC to adopt transportation control measures that would result in sufficient emissions reductions "to attain the NAAQS" or to delay highway projects "to avoid violations of the NAAQS."[67]

Plaintiffs responded that the Contingency Plan was necessary to ensure that the region achieved its air quality goals, and that although the measures to be undertaken to bring the region into compliance were described only in general terms, it nevertheless committed the defendants to take specific actions and therefore the decision in *Wilder* was not controlling.[68] The plaintiffs pointed out that the language in the Contingency Plan promising to "bring the region back within the RFP line" was more similar to the provision involved in the case of *Atlantic Terminal Urban Renewal Coalition v. Department of Environmental Protection,* in which the City of New York promised to implement mitigation measures to attain the applicable CO standards if the project environmental impact statement (EIS) identified an air quality violation. The district court had ruled there that the relevant language did more than merely restate the goal of complying with ambient air quality standards and thus committed the city to act.[69] Therefore, plaintiffs concluded, by analogy the Transportation Contingency Plan was indeed an emission standard or limitation and could form the basis of a citizen suit under the Clean Air Act.

MTC, however, insisted that there were no specific standards for implementing individual TCMs and that the interpretation of what it would take to "bring the region back within the RFP line" was again a matter within its administrative discretion. The MTC asserted that because the 1987 deadline had passed, it was not clear what would satisfy RFP anyway, and therefore the whole issue should be left up to the EPA, which, MTC claimed, had already endorsed its approach. Throughout this litigation, the MTC insisted that it was up to the EPA, not the federal courts, to determine whether it was complying with the Clean Air Act.

Finally, the MTC also maintained that the 6-month deadline for adopting additional TCMs was not a "specific, non-discretionary, clear-cut requirement" of the Bay Area Plan, enforceable by citizen suit. The 6-month schedule was an administrative schedule, they argued, not a statutory deadline mandated by the Clean Air Act, and that it should be interpreted with a "reasonable degree of flexibility."[70]

Reviewing Projects for Delay

As to the plaintiffs' charge that MTC had failed to delay any prospective transportation projects once it became clear that RFP was not going to be met, the defendants denied that there was any such mandate. They pointed out that the Bay Area Plan specifically stated that MTC considered unrealistic the EPA's requirement that SIPs contain a list of transportation projects to be delayed if RFP was not met. The agency explained it could not know in advance which projects might be under way in the future and so, realistically, it could not provide such a list. Instead, the Plan provided for developing a process to identify potentially polluting projects that would be considered for delay and to select criteria for choosing which projects would be delayed (see Appendix B). By approving the Bay Area Plan, the MTC again argued, the EPA had implicitly endorsed their approach; CBE and the Sierra Club should have protested to the EPA if they did not believe this was adequate, not ask the court to, in effect, rewrite the SIP.

Finally, the MTC argued that CBE and the Sierra Club were attempting to rush the court into issuing an improper and unnecessary order, which would unduly hinder MTC's efforts to improve the air quality in the Bay Area. The agency also reiterated its contention that the entire lawsuit was premature, because the state had been ordered to submit a revised SIP anyway, which, when completed, would address the plaintiffs' concerns.

The MTC contended it had been diligently fulfilling its duties in dealing with a very complex and difficult problem in cooperation with the EPA and the other responsible agencies, and that the plaintiffs were improperly seeking to have the court order it to modify the existing SIP based solely on their own notions of proper environmental policy.[71] The MTC felt that granting the plaintiffs' request would interfere with its efforts to both adopt and implement effective control measures for improving air quality and prepare a revised SIP adequate to attain the NAAQS. The agency urged the court to find it in compliance, especially inasmuch as it would be preparing a new SIP anyway, which would contain a revised definition of RFP to comport with the post-1987 attainment standards then being formulated by Congress.

Court Ruling

In its ruling, the court agreed that the Transportation Contingency Plan provisions did represent specific and mandatory commitments and therefore the plaintiffs had standing to enforce them. The MTC had to adopt additional TCMs, even though the Plan did not specify exactly which ones to enact, and the 6-month deadline constituted a "timetable for compliance" that was also enforceable. The MTC would likewise have to consider delaying any highway projects that could have potentially significant adverse effects on air quality. The court held that the process to delay projects was a specific strategy for achieving cleaner air.

The court concluded that the language in the Plan was more like that in *Atlantic Terminal.* Although it did not mandate any particular actions, but provided only a process for discretionary review, the court still held that the MTC's commitment to undertake that process was fully enforceable. In addition, the court rejected the defendants' contention that the provision was unenforceable because the MTC had discretion to decide what would constitute achieving RFP. Once it became clear that RFP had not been met, the court ruled, the MTC was obliged to delay highway projects and adopt additional TCMs.

MTC pleaded, however, that there had been no delay in implementing the transportation contingency measures, and that as far as it was concerned, RFP had been met for the transportation sector for the years 1983 through 1986. The agency explained that it is difficult to quantify emissions reductions from TCMs, so the Plan had adopted an implementation schedule that was used to measure RFP instead of an "objective, straight-line" test.[72] The MTC believed it had met the implementation schedule for TCMs and that they were generally effective in reducing CO emissions. Because it had adopted the required TCMs, the MTC maintained that RFP had been achieved for CO.[73] Although a non-RFP finding for CO was made for 1987, the MTC contended that it had no legal obligation to implement the Contingency Plan until at least July 1989, when the EPA had formally requested that it do so. By that time, the MTC explained, it had already initiated the process on its own.[74] With respect to ozone, TCMs were expected to contribute only marginally to reducing HCs and so they were not even included in the SIP attainment demonstration. Thus, the MTC asserted, it had no liability for ozone emissions.[75]

The court, however, agreed with the plaintiffs that the MTC had failed to carry out its part of the Contingency Plan, because it clearly had neither considered any projects for delay nor adopted any additional TCMs. First, it held that RFP for CO had in fact not been met for the years 1983 through 1987 and that (as noted earlier) the region had failed to achieve ozone standards for the years 1986 and 1987. This, the court concluded, triggered the contingency provisions independent of any EPA enforcement actions,[76] Second, it directed the MTC to adopt sufficient TCMs within 6 months to bring the region back within the RFP line. In addition, the court specified a timetable for completing any hearings for delaying projects and ordered defendants to file progress reports with the court every 90 days.

The MTC began to consider a number of possible transportation controls, including bridge toll hikes, gas tax increases, highway tolls, prohibiting employers from offering free employee parking, and even gas rationing. MTC officials worried that local measures would not be sufficient and urged the state legislature to approve yearly automobile inspections. The outcome of MTC's effort to devise and adopt adequate TCMs to meet the RFP targets is taken up in Chapter 4. Meanwhile, those issues left unresolved by the court's March 5th Order were decided by the court in a separate ruling in May 1990. Among them was whether the MTC had properly carried out the transportation conformity assessments required by Appendix H of the Bay Area Plan. The MTC maintained that the provisions in the Conformity Assessment were not enforceable in a citizen suit and that even if they were, the agency had fulfilled them in a satisfactory manner.

Failure to Implement Conformity Review Procedures

As with the Contingency Plan, the MTC also contended that the Conformity Assessment was not an emissions standard or limitation. Here again, though, the court held that the plaintiffs had standing to enforce these provisions.[77] As the court explained, under the Clean Air Act an emission standard or limitation included any "indirect source review requirements" in a SIP, and the Conformity Assessment was such a requirement because highway projects were indirect sources of pollution because they generate "additional mobile sources of pollution."[78] Therefore, plaintiffs could challenge its implementation just like any other "substantive emission requirements" in the Plan. This marked the first court decision to hold conformity provisions to be an enforceable part of a SIP.

On the question of compliance, plaintiffs alleged that in addition to evaluating TCM implementation, the MTC had specifically committed itself to determining the air quality impacts from projects in the RTP amendments and the TIP, but that the agency had not in fact assessed the projects in the TIP for conformity with the Bay Area Plan before granting them approval. The MTC claimed that it had conducted TIP assessments, even if the plaintiffs did not like the way they did them, explaining that because a "complete air quality analysis" of each project in the TIP would have been "too expensive, time consuming and have little credibility," the agency had adopted a more general approach.[79] The MTC had placed projects in one of three categories based on the nature of the projects: "neutral," "beneficial" or "potential negative effect." Most projects were grouped as either "N" (neutral) or "B" (beneficial). For instance, these included "Rehabilitation Projects" (major damage restoration and highway planning) and "Operational Improvements" (bus/car pool lanes, passing lanes, improving signalization, etc.). Only projects involving new highway construction were considered to have potential negative effects and given a "D" designation.[80]

Plaintiffs, however, insisted that EPA regulations required the conformity assessments to "offset any increase in air pollution concentrations that are expected to result from emission increases due to projected growth of population, individual activity, motor vehicle traffic or other factors."[81] This, according to the plaintiffs, demanded that a quantitative analysis be conducted, rather than the "qualitative" determination adopted by the MTC. The MTC countered that the language in the Plan was never intended to require such an approach and that reading anything more into the Plan would not be enforcing a "specific strategy" but rather improperly revising an existing SIP.

The court agreed that the MTC's cursory approach did not provide a realistic assessment of the impacts of highway projects on air quality inasmuch as the MTC did not independently evaluate potential environmental impacts and even allowed some projects to be classified "without reference to any outside environmental study at all." Also, the court held that the MTC's intentions were irrelevant because they were contrary to the "plain terms" of the Plan, and anyway, the state could not "write around" the requirements of the Clean Air Act. Moreover, the court pointed out that there was no category for projects that "will adversely affect emissions" as required by Appendix H, terming the "potential negative effects" category "particularly useless." Although projects in this "D" category might eliminate or reduce traffic congestion, they could also encourage future growth in travel, which in time could "override the initial benefits." All this made the conformity procedures a "pointless exercise with no potential for meaningful application," according to the court.[82] The court concluded that the MTC would have to adopt a proper conformity assessment even if that would mean additional costs to the agency.

Although the court held that the MTC's "qualitative" evaluation was not adequate to satisfy the legal requirements, it deferred a ruling at that time on the precise remedy. Instead, the court ordered the MTC to submit a proposal within 30 days to implement new consistency provisions with respect to the TIP and to file a supplemental brief regarding its RTP conformity procedures.[83] The agency subsequently agreed to modify its procedures for both the RTP and the TIP assessments. The MTC submitted its revised procedures to the court for review in July and the plaintiffs filed a number of objections to them. The details of the MTC's proposed new conformity assessment procedures and the plaintiffs' response to them are taken up in the next chapter.

Summary

The district court made several important rulings in this initial liability phase of the case. First, it ruled that the citizens' organizations could force the MTC and other regional planning agencies to adhere to the commitments made in the Bay Area Plan. The defendants had to adopt the stationary source controls that had been listed in the plan but never carried out. In addition, because the region had not made reasonable further progress toward meeting the clean air goals, as committed to in the SIP, they also had to devise and implement additional measures covering both stationary and vehicular sources of air pollution.

Second, because the defendants had failed to reduce emissions levels as promised in the Bay Area Plan, the court also ruled that the contingency TCMs had to demonstrate sufficient measurable reductions in emissions to erase the shortfall. Finally, the court held that under the Clean Air Act and the Bay Area Plan, the MTC had to adopt a quantitative assessment process for evaluating the conformity of transportation plans, programs, and projects to the SIP. This marked the first time a court had specifically enforced the contingency and conformity provisions in a SIP.

The court's decision adhered to existing federal law in holding that the defendants could not be held liable simply because their efforts had not resulted in the region's achieving the federal air quality standards. On the other hand, the court took a fairly aggressive approach in finding that the provisions in the Bay Area Plan were specific enough to constitute enforceable commitments under the Clean Air Act. In its decision, the court succinctly captured the different views of the purpose and effect of the Bay Area Plan held by the defendants on the one hand and by the court and plaintiffs on the other:

> To recap, defendants paint the Bay Area Plan as an advisory document containing "suggestions" and "recommendations" but no commitments to implement any particular strategy by any definite time. As discussed above, such a view is contradicted by the plain language of the Plan, inconsistent with the case precedent, and contrary to the Act itself.[84]

The decision lends weight to efforts by both citizens groups and federal regulators such as EPA to ensure that local agencies keep their planning commitments.

The court also took a fairly hard line in not deferring to the EPA's administrative process but instead holding that the defendants were obligated to implement the contingency measures as soon as the RFP report established that the region had fallen "off the RFP line." With the transportation contingency measures, it was fairly easy to order the MTC to take action, because the plan specifically called both for adopting "sufficient TCMs" to satisfy RFP and for delaying any projects with significant environmental consequences, as required by the EPA. The liability of the ARB and District proved more difficult to specify, because the plaintiffs initially asked the court to direct the defendants to adopt additional measures needed to attain the NAAQS and this could have been interpreted as merely reiterating the statutory requirements. The Plan, however, did contain specific numerical targets for reducing emissions from ozone, namely by 85 tons per day. That, according to the court, constituted a "specific strategy" for achieving the NAAQS that could be enforced.

Judge Henderson's decision drew strong editorial rebukes from the *San Francisco Examiner* and the *Wall Street Journal,* each accusing the judge of overstepping his judicial authority and requiring the government to take impractical steps to curb pollution that might actually increase congestion.[85] Bill Curtiss, the lead attorney for the Sierra Club Legal Defense Fund, responded to these charges of judicial activism by insisting that the court was merely enforcing a plan prepared and approved by both the state and the EPA, as it was authorized to do by Congress.[86]

Given the national interest sparked by this decision, what, on a practical level, can be said to be its impact? By holding that the plaintiffs could enforce specific contingency provisions and conformity procedures, it certainly lends support to efforts by environmentalists and federal regulators to force states to clean up their air. An opinion by a federal district judge, though, has limited value as a precedent value. Although it is binding between the parties to the case, and it can have some persuasive force in other similar litigation, other state and federal courts are not bound to follow it. It can, however, affect the direction of federal regulatory policy. For the EPA, it confirmed that SIP commitments are enforceable, which will add some leverage to their rule making and oversight responsibilities. Perhaps buoyed by the success in the Bay Area lawsuit, shortly after passage of the 1990 Clean Air Act Amendments, several national environmental organizations notified a number of metropolitan planning organizations and most state governments demanding a larger role in the planning process and warned them to comply with the new transportation requirements or face lawsuits. More immediately, the decision certainly did have a dramatic impact on transportation planning in the Bay Area, as the following chapters illustrate. It forced planners in various state and regional agencies to seriously consider the relationship between transportation and air quality. It also pushed them to develop new analytical techniques.

In one sense, the court's rulings can be viewed as relatively narrow, because they are based on interpreting the language in the Bay Area Plan, and other SIPs will of course be different. Planners in other areas may be able to avoid liability simply by being more careful in their choice of language in their plans, by more explicitly stating that certain implementation measures are merely suggestions for future consideration, or by clearly reserving greater discretion to choose among various policy options. On the other hand, as the plaintiffs pointed out in this case, the EPA requires plans to contain enforceable commitments and may refuse to approve plans it considers too vague. Admittedly, it is not possible to predict how other courts may react in similar circumstances; however, if other courts do follow the lead of Judge Henderson, they may be inclined to construe plan language more strictly against the responsible agencies in order to comply with the congressional mandate to improve air quality as soon as reasonably possible.

In the broader context, the judge's ruling that clean air plans are binding reflects the growing trend discussed earlier to treat planning documents more seriously, to expect the government to not only follow through on what it states it intends to do regarding a given problem but also to state exactly how it will measure the success or failure of its actions and what it will do in the event that its programs do not achieve their intended result. In that sense, the court's decision is perhaps more important for its general approach than the specific holdings. Courts can and should be expected to continue to defer to agencies' administrative discretion in areas of their expertise. They may be more inclined, however, once agencies commit themselves on paper to a particular course of action, to require that they follow though and to not tolerate endless delays. Indeed, as pointed out in Chapter 1, the 1990 amendments require many areas to adopt "emissions budgets," which are likely to be quite similar to the RFP targets upheld in this case. Contingency measures must take effect automatically if the budgets are not met, and eventually mandatory penalties may be imposed. Given Congress's desire to "get tough" on air pollution, it is reasonable to assume that courts will be at least as stringent with these new requirements as this court was in construing the contingency measures in the Bay Area Plan. Planners may find that if they say it in the plan, the courts will hold them to it.

As the case went on, the emphasis shifted away from the stationary source sector to the MTC's obligations for reducing transportation-related emissions. As mentioned earlier, the MTC was not only jointly liable for the 85 tpd reductions in HC but also solely responsible for the entire 393 tpd reduction in CO emissions. The agency was also in charge of seeing that plans for the transportation system conformed to the air cleanup plan. These initial rulings necessitated the adoption of new methods of modeling the air quality impacts of alternative transportation systems. This new methodology would be used both as the basis for the MTC's proposed new conformity assessments and in evaluating the effectiveness of the various TCMs that the MTC began to consider to close the gap between the expected and actual air pollution levels. Both sides retained outside experts, to assess the various proposals, and proceeded to argue about the merits of different modeling approaches. Because the court had ruled that the MTC had to improve its analytical procedures, it became involved in evaluating the details of that methodology, an untypical role for the courts, which tend to defer to administrative agencies on technical issues.

From here on, the lawsuit turned mostly into a battle over new highway construction, and the lawyers for the Sierra Club took on the primary responsibility for presenting the plaintiffs' case. Publicly, officials for the MTC maintained that the lawsuit should never have been brought and was simply a trivial argument over what they saw as "a few missed deadlines." The real purpose of the lawsuit, in their view, was to halt or delay needed highway improvements for political purposes. The commission saw its air quality responsibilities as being in conflict with its mission to improve mobility. It saw the environmentalists as antihighway crusaders interfering with the mandate of elected public officials. MTC officials insisted, though, that despite Judge Henderson's ruling, highway projects would not be affected.[87] For their part, the plaintiffs felt that the MTC was "freeway happy" and would approve any project without regard to its air quality impact rather than risk losing federal highway funds. The MTC believed that, on the contrary, their projects helped relieve traffic congestion and therefore reduced pollution. Plaintiffs countered that new freeways simply encouraged more unwanted growth and that soon the roads would be just as crowded as before, but at much higher volumes, producing even more air pollution. The dispute crystallized around the parties' two conflicting views of the impact of traffic on air quality and around their assessments of the capabilities of traffic forecasting and travel modeling.

The following chapters describe how the MTC attempted to comply with the court's orders and how the court interpreted those actions. The experience is particularly instructive, especially in light of the 1990 amendments to the Clean Air Act, which mandate that all states begin doing many of the things

Judge Henderson concluded California had obligated itself to do under the 1977 act in the Bay Area Plan. The next chapter looks at the details of the defendants' new modeling approach and how the parties' differing viewpoints regarding the defendants' air quality responsibilities influenced their dispute over the Conformity Assessment. Among the disputed issues between the parties were (a) how to define conformity with respect to plans, programs, and individual highway projects; (b) the proper assumptions to be used to estimate emissions from the traffic models; and (c) whether the assessment process adequately considered the impact that transportation plans would have on the size and distribution of regional growth. Following that, the question of the adequacy of the TCM program is taken up in Chapter 4.

Notes

1. The two cases, *Citizens for a Better Environment, et al. v. Deukmejian, et al.,* Case No. C89 2044 TEH, and *Sierra Club v. Metropolitan Transportation Commission, et al.,* Case No. C89 2064 TEH, were consolidated for all purposes under the title of the first case by order of court dated August 8, 1989.

2. The San Francisco Bay Area Air Quality Planning District includes the City and County of San Francisco, and San Mateo, Marin, Napa, Contra Costa, Alameda, and Santa Clara counties and portions of Sonoma and Solano counties.

3. The federal standards for ozone were exceeded in the Bay Area on 15 days during 1979. Levels measured in the Bay Area (primarily in the Santa Clara Valley) reached 19 pphm, compared to the 12 pphm federal standard. The federal 1-hour CO standard of 35 ppm was not exceeded; however, the 8-hour standard of 9 ppm was exceeded on 21 days in 1979. 1982 Bay Area Air Quality Plan (BAAQP), at 2.

4. In July 1979, the EPA issued an interpretative rule that the construction ban in Section 110(a)(2)(I) would apply to any area not yet covered by an approved Part D SIP. 44 Fed. Reg. 38,471, 38,473 (July 2, 1979) (codified at 40 C.F.R. 52. 24(a) (1994)). See discussion in chapter 1, note 47 supra.

5. The EPA approved the extension for the Bay Area. 47 Fed. Reg. 50,864 (November 10, 1982) (codified at 40 C.F.R. § 52. 222(d)(9) (1994)).

6. 47 Fed. Reg. 11,866 (March 19, 1982). The EPA conditionally approved the resource and extension requirements for CO and ozone in the 1979 nonattainment plan, provided that the state submit implementation schedules and resource commitments for the TCMs in the plan; however, it disapproved the I/M portions of the plan, which resulted in an overall disapproval. The EPA had also originally proposed to disapprove the plan because it did not contain legally enforceable New Source Review (NSR) rules; however, the state had eventually complied with the EPA's conditions and this objection had been removed. See 45 Fed. Reg. 21,282 (April 1, 1980); 45 Fed. Reg. 41,983 (June 23, 1980); 45 Fed. Reg. 48,164 (July 18, 1980).

7. Cal. Health & Safety Code § 39602 (West 1986).

8. MTC, ABAG, and the District were colead agencies for preparing the revised nonattainment plan for the Basin pursuant to Part D of the Clean Air Act. MTC was responsible for those

plan elements designed to reduce vehicular sources of air pollution by (a) encouraging alternate modes of travel, (b) reducing the quantity of travel, and (c) shifting the spatial and temporal distribution of vehicular emissions. ABAG was responsible for any plan elements related to regional development policy. The District was responsible for any plan elements and control measures dealing primarily with nonvehicular sources of air pollution.

9. See Cal. Health & Safety Code, Division 26, Part 3, Chapter 4, §§ 40200 et seq. (West 1986 & Supp. 1995).

10. See Cal. Gov't Code §§ 66500 et seq. (West 1983 & Supp. 1995).

11. ABAG was established by a joint powers agreement. See Cal. Gov't Code, Title I, Div. 7, Ch. 5, Art. 1, § § 6500 et seq. (West 1980 & Supp. 1995).

12. Cal. Gov't Code §§ 53098 (West 1983).

13. Cal. Gov't Code § 65584(a) (West Supp. 1995). The section provides that each Council of Governments (COG) is to determine its regional share of existing and projected statewide housing needs.

14. The state submitted a draft plan on July 14, 1982. In February 1983, the EPA proposed to conditionally approve the SIP revisions for ozone and CO for the Bay Area. The final plan, submitted on February 4, 1983, included a commitment to report on progress in developing and implementing the RFP process for CO in the annual reports and corrected other minor deficiencies. It also included resource commitments to implement the 10 TCMS, removing a condition of approval of the 1979 Nonattainment Plan (BAAQP, Appendix I). On the basis of its review, the EPA approved the 1982 Bay Area Air Quality Plan (except for the I/M element) under Part D of the Clean Air Act and incorporated it into the state SIP, effective January 27, 1994. 48 Fed. Reg. 57,130 (December 28, 1983) (codified at 40 C.F.R. § 52. 220(c)(135) (1994)).

15. The state legislature eventually approved the I/M Program on September 10, 1982, and submitted it to the EPA for approval. The EPA proposed to retain its disapproval of the I/M portion of the nonattainment plan for the Bay Area due to lack of an adequate implementation schedule. 48 Fed. Reg. 5074 (February 3, 1983). Following certain revisions, the EPA conditionally approved the 1979 I/M program SIP requirements and agreed to lift the construction moratorium. 48 Fed. Reg. 53,114 (November 25, 1983).

16. BAAQP, at 8.

17. Id. at 53, Table 3.

18. LIRAQ is a gridded, single-layer photochemical model that simulates changing concentrations of reactive and nonreactive pollutants over space and time and was designed to deal with the complex topography in the Bay Area. It is especially useful for evaluating pollution control strategies because it can be used to determine the relative role played by various sources or species of air pollution in regional air quality. Using this information, different strategies can be simulated to measure their effects. Id. at 62-64.

19. Id. at 94. The formula to calculate the percentage reduction needed to lower the design value to the federal standard would be (14.4-12)pphm/14.4 pphm = 0.166 or approximately 17%.

20. Id. at 110-116, especially Table 27 and Figure 13. Implementation of all recommended control measures was expected to produce a new 1987 design value of 11 pphm, which would be below the 12 pphm federal standard. See 48 Fed. Reg. 5074, 5078, Table 2 (February 3, 1983).

21. Id. at 42. HC emissions from vehicles subject to the I/M program were estimated at 116 tpd in 1987. The smog check program was expected to be 25% effective in reducing emissions. Therefore, 116 tpd × 0. 25 = 29 tpd.

22. Id. at 42. Carbon monoxide emissions from vehicles subject to the I/M program were estimated to be 1,468 tpd in 1987. Again, the smog check program was expected to be 25% effective in reducing emissions. Therefore, 1,468 tpd × 0. 25 = 367 tpd.

23. The Bay Area Plan evaluated 15 proposed TCMs; only 10 were recommended for inclusion and the other 5 were considered not reasonably available. Id. at 104 and Appendix B.

24. Id. at 102-104. The reasons given included (a) the vehicle fleet would be much cleaner by 1987 so travel reductions would have less impact; (b) TCMs have only minor impact on travel decisions; (c) Bay Area residents already used alternative transit modes to a large extent; and (d) TCMs reduce HCs and NOx simultaneously and so have little impact on ozone levels.

25. Id. at 132-135. Other contingency measures for San Jose included (a) implementation of more stringent CO exhaust emission standards on light-duty vehicles, (b) a master synchronized signal control system for downtown San Jose, and (c) a transportation system management plan and parking policies for downtown San Jose. Limiting commercial and retail development in the central business district was considered counter to the city's economic development goals and not likely to be effective because background CO accounted for the major portion of total pollution concentration in the area.

26. Id. at 137-139, especially Table 31 and Figure 21.

27. 48 Fed. Reg. 5074, 5080 (February 3, 1983).

28. BAAQP, at 138-139, Table 31 and Figure 21. The program to evaluate RFP included

1. an annual program of traffic counts at key intersections in downtown San Jose,
2. a procedure to assess the effectiveness and progress of the Commute Transport Program,
3. tracking vehicle turnover and deterioration trends,
4. tracking effectiveness of motor vehicle inspection and maintenance,
5. tracking ambient CO concentration level, and
6. implementation of Advisory Review and Conformity Assessment.

Id. at 135-137.

29. Cal. Gov't Code §§ 66508-13 (West 1983 & Supp. 1995).

30. Approval of 1982 Ozone and Carbon Monoxide Plan Revisions for Areas Needing an Attainment Date Extension, 46 Fed. Reg. 7182, 7187 subd. I.D. 7 (January 22, 1981); see *Delaney v. E.P.A.,* 898 F.2d 687, 693 (9th Cir. 1990).

31. EPA letters to the MTC, dated May 28, 1988, June 13, 1988, and February 24, 1989.

32. Final Policy—Criteria for Approval of The 1982 Plan Revisions, 46 Fed. Reg. 7187, subd. I.D. 8 (January 22, 1981).

33. The circumstances that would make it necessary to review and develop stationary source contingency measures included

1. the I/M Program being less effective than 29 tons/day,
2. one or more of the recommended measures being less effective than analysis indicated,
3. delays in adoption of regulations or delays in compliance,
4. multiple exemptions changing reduction estimates, and
5. air monitoring data showing no air quality improvement.

BAAQP, at 101.

34. Id. at 106-107.

35. EPA letter to the MTC, dated February 24, 1989. See discussion in chapter 1, note 77 supra.

36. Section 304(b) of the Clean Air Act requires that 60 days prior to the filing of a citizens' suit in federal court under Section 304(a) of the act, the alleged violator, the EPA, and the state in which the alleged violations occurred be given notice of the action. 42 U.S.C.A. § 7604(b) (West 1983 & Supp. 1995).

37. *Oakland Tribune,* August 28, 1989, A-7; *San Jose Mercury News,* June 14, 1989, 7B.

38. Plaintiff CBE filed a motion for partial summary judgment (July 13, 1989) against all five state and local agency defendants on the issue of liability for failure to adopt the required ozone reduction measures. Plaintiff Sierra Club filed separate motions for summary judgment against the District and the ARB (August 2, 1989) for failure to adopt all the stationary source controls, and against the MTC and the ARB (August 17, 1989) for failure to carry out the contingency plan and to assess the air quality impact of transportation plans and programs. Hearings on these motions were held on September 18 and 19, 1989, and the court issued a written partial ruling on March 5, 1990.

39. *Citizens for a Better Environment v. Deukmejian* (CBE I), 731 F. Supp. 1448, 1452, 31 Environment Reporter Cases (ERC) 1213, 1216 (N. D. Cal 1990). See also *National Resources Defense Council, Inc. (NRDC) v. New York State Dep't of Environmental Conservation,* 668 F. Supp. 848, 852 (S.D.N.Y. 1987); *American Lung Association of New Jersey v. Kean,* 670 F. Supp. 1285, 1289 (D.N.J. 1987) *aff'd,* 871 F.2d 319, 322 (1989).

40. CBE I, 731 F. Supp. at 1456, 31 ERC at 1219-1220.

41. The District and the ARB subsequently complied and adopted the following measures: District Regulation 8, Rule 42 (governing large commercial bread bakeries) & Rule 49 (governing aerosol paint); ARB regulations controlling emissions from utility or reciprocating engines and governing consumer solvents. See *Dunn-Edwards Corporation v. Bay Area Air Quality Management District,* 9 Cal. App. 4th 644, 11 Cal. Rptr. 2d 850 (1992) (application of California Environmental Quality Act (CEQA) to District amendments to Regulation 8, Rules 3 & 48 (architectural coatings) held not preempted by decision in *Citizens for a Better Environment*).

42. Section 304(a) provides in part (1990 amendments in italics):

> Except as provided in subsection (b) of this section, any person may commence a civil action on his own behalf—
>
> (1) against any person (including (i) the United States, and (ii) any other governmental instrumentality or agency to the extent permitted by the Eleventh Amendment to the Constitution) who is alleged to *have violated (if there is evidence that the alleged violation has been repeated) or to* be in violation of (A) an emission standard or limitation under this chapter or (B) an order issued by the Administrator or a State with respect to such a standard or limitation, . . .
>
> (2) . . .
>
> The district courts shall have jurisdiction, without regard to the amount in controversy or the citizenship of the parties, to enforce such an emission standard or limitation, or such order, or to order the Administrator to perform such act or duty, as the case may be, *and to apply any appropriate civil penalties (except for actions under paragraph (2)).*
>
> 42 U.S.C.A. § 7604(a) (West 1995).

43. *Wilder v. Thomas,* 854 F.2d 605, 613-16 (2d Cir. 1988) (a citizen suit may not be brought alleging violation of a NAAQS or a general policy expressed in a SIP); *Riverside Cement Co. v. EPA,* 843 F.2d 1246 (9th Cir. 1988).

44. *Plan for Arcadia, Inc. v. Anita Associates,* 379 F. Supp. 311, 324, *aff'd,* 501 F.2d 390 (9th Cir. 1974), *cert. denied,* 419 U.S. 1034 (1974).

45. 42 U.S.C.A. § 7604(a)(1) (West 1995). See note 42 supra for full text.

46. 42 U.S.C.A. § 7604(f) (West 1995). See also *Oregon Environmental Council v. Department of Environmental Quality,* 775 F. Supp. 361 (D. Or. 1991).

47. 42 U.S.C.A. § 7607(b)(1) (West 1995).

48. See *American Lung Association of New Jersey v. Kean,* 670 F. Supp. 1285 (D.N.J. 1987) *aff'd,* 871 F.2d 319, 322 (1989) (district court had subject matter jurisdiction over suit alleging that state violated its own implementation plan for ozone pollution control strategies under express language of the Clean Air Act); *Friends of the Earth v. Carey,* 535 F.2d 165, 169 (2nd Cir. 1976), *cert. denied,* 434 U.S. 902, 98 S. Ct. 296, 54 L. Ed. 2d 188 (1977); *NRDC* v. *New York,* 668 F. Supp. at 850-51.

49. American Lung Ass'n, 670 F. Supp. at 1292. See discussion in chapter 1, note 74 supra.

50. EPA's Response to Sierra Club's Motion for Summary Judgment Against MTC and ARB, September 1, 1989, at 4-5.

51. These were identified as covering Architectural Coatings, Ship Barge Tanker RR Load, General Solvents & Surface Coating Operations, Pleasure Boats, New Source Review, New Service Stations, Lawn Mowers, Offroad Motorcycles, Wineries, Marine Vessel Gas Freeing, and Marine Lightering Retrofit. BAAQP, at 99, Table 22.

52. Id. at 150.

53. CBE I, 731 F. Supp. at 1459, 31 ERC at 1222 (quoting *Action for Rational Transit* v. *West Side Highway Project,* 699 F.2d 614, 616 (2d Cir. 1983)).

54. *Citizens for a Better Environment v. Deukmejian* (CBE II), 746 F. Supp. 976, 980, 32 ERC 1136 (N.D. Cal. 1990). The motion for reconsideration was heard on May 31, 1990. Following additional submissions at the request of the court, the court issued a written order on August 28, 1990.

55. See American Lung Ass'n, 670 F. Supp. at 1292 (" . . . compliance with the ozone standards themselves is an issue logically once-removed from the question of SIP compliance").

56. MTC's Memorandum of Points and Authorities in Opposition to Motion by Plaintiff Sierra Club for Summary Judgment, September 5, 1989, at 4.

57. CBE II, 746 F. Supp. at 981, n. 8, 32 ERC at 1140, n. 8. This figure was based on different assumptions from those used to prepare the 1982 Bay Area Air Quality Plan. On the basis of the estimated "uncontrolled" emissions of 515 tpd, the emissions with controls in 1987 would have been 473 tpd. The court ruled that the reason for the shortfall was immaterial because liability in a citizen's suit "turns solely on the question of actual compliance." Id. See chapter 3 for a full discussion of the conflicting emissions estimates.

58. Association of Bay Area Governments, *Bay Area Air Quality 1987 RFP Report,* November 1988, 16 [hereinafter 1987 RFP Report].

59. Declaration of Edward Miller, June 21, 1990, Ex. F. See also, California I/M Review Committee, *Evaluation of the California Smog Check Program: Second Report to the Legislature,* May 1989.

60. CBE II, 746 F. Supp. at 981, 32 ERC at 1140.

61. Id. at 982, n. 9, 32 ERC at 1140, n. 9.

62. Association of Bay Area Governments, *Bay Area Air Quality RFP Reports: A Retrospective Assessment of 1985 and 1986,* July 1988, 16 [hereinafter 1988 Retrospective]. The 1988 Retrospective indicated CO shortfalls in both years. Reductions of only 107 tpd were achieved in 1985 from the I/M program and 15 to 22 tpd from the 10 TCMs, according to the report. For 1986, the reductions amounted to only 144 tpd from the I/M program and 14 to 19 tpd from the TCMs.

63. Association of Bay Area Governments, et al., *Air Quality Planning in the Bay Area: 1988 Status Report,* March 1988 [hereinafter 1988 Status Report]. According to the 1988 Status Report, the I/M program was only 12.3% effective for HCs in 1986 and only 9.8% effective for CO emissions (1,468 × .098 = 144 tpd), instead of the anticipated 25%. Short-term improvements to 26.9% for HC and 16.3% for CO were predicted, along with long-term improvements to 39.7% and 25.2%, respectively. California I/M Review Committee, *Evaluation of the California Smog Check Program: First Report to the Legislature,* April 1987.

64. Metropolitan Transportation Commission, *Bay Area Travel Forecasts: Year 1987 Trips by Mode, Vehicle Emissions, Technical Summary,* August 1990, 14. The report estimated 1987 daily vehicle trips to be 17% higher than forecast in the 1982 Bay Area Plan.

65. Sierra Club's Reply Memorandum in Support of Motion for Summary Judgment against MTC and ARB, September 11, 1989, at 23.

66. *Wilder* v. *Thomas,* 854 F.2d 605, 608-10 (2d Cir. 1988), *cert. denied,* 489 U.S. 1053, 109 S. Ct. 1314, 103 L. Ed. 2d 583 (1989) (elimination of all CO hot spots by end of 1987, as required by national ambient air quality standards and state implementation plan, was not a condition or requirement relating to transportation control measure within meaning of statute because NAAQS for CO were not specific provisions of SIP requiring all CO hot spots to be eliminated by end of 1987). See also *League to Save Lake Tahoe, Inc., v. Trounday,* 598 F.2d 1164, 1173 (9th Cir. 1971), *cert. denied,* 444 U.S. 943, 100 S. Ct. 299, 62 L. Ed. 2d 310 (1979).

67. EPA's Response to Sierra Club's Motion for Summary Judgment Against the MTC and ARB, September 1, 1989, at 2.

68. Plaintiffs also pointed out that MTC had responded to EPA criticism of an earlier version of the SIP as being too vague by clarifying that the Contingency Plan was an enforceable commitment to adopt TCMs within 6 months of a non-RFP decision.

69. *Atlantic Terminal Urban Renewal Area Coalition v. New York City Depart. of Environmental Protection,* 697 F. Supp. 157, 162 (S.D.N.Y. 1988).

70. Defendant MTC's Memorandum of Points and Authorities in Opposition to Motion by Plaintiff Sierra Club for Summary Judgment, September 5, 1989, at 13-14 [hereinafter MTC's Memorandum].

71. The proposed order would have required the MTC to (a) within 120 days, adopt, implement, and enforce additional TCMs to attain NAAQS; (b) within 90 days, review each transportation project included in the RTP or TIP since January 1983, or which is proposed for inclusion therein; (c) within 120 days, circulate a list of such projects, a quantification of the effects of such projects, criteria to be used for delay of such projects, and an identification of the projects to be delayed such that projects that are not delayed will not increase emission levels necessary to attain or maintain the NAAQS; and (d) within 180 days, take final action to withdraw approval of projects to be delayed and thereafter not approve any project without determining that its effect on air quality, emissions, VMT, or vehicle trips will not delay the attainment of NAAQS or cause a violation of the standards. The MTC insisted that this schedule was totally infeasible, given administrative and technical constraints.

72. MTC's Memorandum, supra note 70, at 4. See BAAQP, at 110-116.

73. In fact, the first RFP report for the Basin (due July 1983), which was not filed until January 1985, did not demonstrate sufficient reductions to attain the CO ambient air quality standard by December 31, 1987. The RFP report for 1984 (due July 1985), which was finally submitted in June 1986, also did not demonstrate RFP for CO. No complete RFP report for 1985 was ever prepared. In April 1986, the EPA concluded that RFP for CO had not been demonstrated for the year 1983 and directed the MTC to submit additional control measures within 6 months. Sierra Club's Memorandum of Points and Authorities in Support of Motion for Summary Judgment against MTC and ARB, August 17, 1989, at 9-12. MTC contended, however, that the EPA's non-RFP determination for CO had been pro forma, based on a lack of information, and did not require implementation of the Contingency Plan. Moreover, that this "ambiguous" determination had been "superseded" by an agreement between the MTC and the EPA, in December 1986, for the MTC to provide the EPA with a complete RFP report for the year 1985. MTC's Memorandum, supra note 70, at 4-6.

74. MTC's Memorandum, supra note 70, at 16-17. See 1987 RFP Report, supra note 58, at 40, Table 12. According to the 1987 RFP Report, estimated CO emissions reductions for 1987

were only 16 to 20 tpd rather than the 26 tpd predicted in the Plan. See also 1988 Status Report, supra note 63; 1988 Retrospective, supra note 62.

75. TCMs do not affect travel and are ineffective at reducing ozone because they reduce HC and NOx simultaneously. Indeed, reductions in NOx emissions could actually be expected to increase ozone production. Although no credit was taken in the Bay Area Plan for the 10 TCMs toward the projected 85 tons reduction in HCs, the 10 TCMs were expected to reduce HC by another 2.6 tons and NOx by 2.8 tons. See Table 2.1 in the text.

76. CBE I, 731 F. Supp. at 1460, 31 ERC at 1223 ("On the contrary, the Plan is very clear that MTC must activate the contingency plan upon a determination that RFP has not been made. . . . Indeed, the [Clean Air Act] provides for citizen suits precisely to 'circumvent bureaucratic inaction that interferes with the scheduled satisfaction of the federal air quality goals' "). See *Friends of the Earth v. Carey*, 535 F.2d 165, 178 (2nd Cir. 1976), *cert. denied*, 434 U.S. 902 (1976) (holding that the Clean Air Act does not empower either the EPA or the state to delay the approved plan's strategies through negotiation, formal or otherwise, adding: "Negotiations are no substitute for enforcement and for timely compliance with the Plan's mandated strategies").

77. *CBE v. Deukmejian*, 31 ERC 1545 (N.D. Cal. 1990). The court, however, denied CBE's motion for summary judgment against ABAG and the governor, finding that neither ABAG nor the governor was liable for failing to implement the contingency measures contained in the Bay Area Plan.

78. Id. at 1546-1547. See *League to Save Lake Tahoe, Inc. v. Trounday*, 598 F.2d 1164, 133 ERC 1801 (9th Cir. 1979), *cert. denied*, 444 U.S. 943, 133 ERC 1883 (1979) (construing "transportation control measures" in 42 U.S.C. § 7604(f) to include all indirect source review requirements).

79. *CBE v. Deukmejian*, 31 ERC at 1547. The MTC insisted that the small changes in VMT and travel speeds between building or not building most projects would have little effect on air quality.

80. Only 82 of the 208 proposed projects in the TIP were in this category. W. Gibbs, "Clean Air Ruling Hits Road Plans," *San Francisco Examiner*, September 20, 1990. Some of these were later given categorical exemptions or received negative environmental declarations.

81. 40 C.F.R. § 51. 110(a) (1994).

82. *CBE v. Deukmejian*, 31 ERC at 1548.

83. Id. at 1548-1550. MTC argued that it had prepared an "appropriate environmental document," which included a "review of the air quality impacts of amendments to the RTP" but failed to provide the court with a copy of the review or even a summary of the methodology used. Although the court found no evidence that the MTC had complied with the Bay Area Plan requirement to determine the air quality impact of the annual RTP amendments, it also found that plaintiffs had failed to prove that MTC had violated the conformity assessment provisions. Id. at 1549.

84. CBE I, 731 F. Supp. at 1458, 31 ERC 1221-1222.

85. Editorial, "Judicial Coup in Transportation," *San Francisco Examiner*, September 24, 1989; Editorial, "Justices vs. Judges," *Wall Street Journal*, January 12, 1990.

86. Editorial (William S. Curtiss), "Disputes Over Clean Air Plans Get Dirtier," *San Francisco Examiner*, February 24, 1990; Editorial (William S. Curtiss), "Judge Wrongly Blamed in Air-Pollution Ruling," *Wall Street Journal*, March 6, 1990.

87. Diringer, "Environmentalists Win Highway Ruling," *San Francisco Chronicle*, May 11, 1990, A2.

The Metropolitan Transportation Commission's New Conformity Assessment Procedures

Over the years, there has been progress in reducing vehicle emissions through inspection and maintenance programs and improvements in engine technology. Those gains, however, may be lost as the number of vehicles on the road and the overall number of vehicle miles traveled have increased. The Environmental Protection Agency (EPA) reports that between 1983 and 1990, passenger vehicle miles traveled (VMT) increased by 41% due largely to increases in the workforce and the number of jobs moving to the suburbs.[1] At the same time, average vehicle occupancy has declined, meaning more cars on the road and greater congestion along with slower average speeds. These trends make it harder to reach air quality goals unless ways are found to reduce trip numbers and growth in VMT. Compounding the problem is the fact that improvements in mobile source emissions from stricter tailpipe regulations mean that the marginal benefit from additional transportation control measures will be less. Many environmentalists fear that increasing growth in travel will offset any improvements gained from these other measures, and air quality will suffer as a result.

The conformity assessment is an important element in the clean air strategy. Metropolitan planning organizations (MPOs) are required to demonstrate that their transportation plans, programs, and highway projects are consistent with the approved air quality implementation plan. Under the 1977 Clean Air Act, conformity was interpreted to mean that new highway construction should not interfere with adopted transportation control measures. Lately, more attention is being paid to the relationship between the form of the transportation system and air quality. As discussed in Chapter 1, the 1990 act requires automobile-generated pollution to remain consistent with the estimates of future emissions used in developing the State Implementation Plan (SIP). The role of conformity assessments in achieving the National Ambient Air Quality Standards (NAAQS) has thus become an even more important issue and one to which we will return again in the concluding chapter.

The issue of conformity was vigorously contested in the Bay Area lawsuit. The environmental organizations considered it critical to slowing or halting unwanted freeway expansions that, in their view, simply added to pollution problems. They believed that expanding highway capacity led to more dispersed metropolitan development, which only encouraged more pollution-creating travel patterns. The transportation agency planners, by contrast, saw this issue as far less important, especially because they felt that a particular transportation system has little impact on overall pollution levels. They put far greater emphasis on the use of transportation-related and mobile source controls in curbing emissions.

This chapter describes the Metropolitan Transportation Commission's (MTC's) efforts to develop an acceptable methodology for conducting quantitative air quality analyses of transportation plans and programs, in light of the district court's rulings discussed in the previous chapter. The Sierra Club used the opportunity to attack the MTC for not facing up to the air quality consequences of highway expansion. Also, the plaintiffs attempted to stop the MTC from continuing to approve highway construction projects until those new procedures were approved and in place. The MTC resisted these efforts by belittling the negative impacts of new freeways on automobile emissions and emphasizing their congestion-reducing benefits. In the end, the court approved a new conformity procedure that took greater account of the impact of highway planning on travel demand and emissions, although it did not do all that the environmentalists had hoped for. With the adoption of the 1990 amendments, transportation agencies throughout the country may well have to develop procedures similar in scope to those produced by the MTC in this lawsuit.

The Transportation/Land Use Connection

In preparing the 1982 implementation plan, the MTC made certain assumptions about future population growth and expected travel demand. These were factored into the attainment demonstration in the Bay Area Plan. The strategies in the Plan, such as inspection and maintenance programs and transportation control measures (TCMs), were designed to reduce emissions to a level consistent, as best could be determined, with attaining the NAAQS. They represented the agency's best judgment of how to achieve the clean air standards given what was known at the time and what could be reasonably projected. If it turned out that some assumptions were incorrect or the projections were inaccurate, such as if population growth exceeded the Association of Bay Area Governments (ABAG) forecasts, the MTC did not believe that it should be held responsible for that situation. The agency's position was that it had to account for only the emissions reductions called for in the Plan—29 tons per day (tpd) of hydrocarbons (HCs) and 393 tpd of carbon monoxide (CO)—not that it had to achieve any particular level of emissions. In the event the transportation measures proved less effective than assumed, the contingency measures were designed to keep the region within the reasonable further progress (RFP) line. If that still did not result in achieving the NAAQS, then the EPA or Congress could require the region to submit a new attainment plan.

As a consequence, the MTC saw the conformity question as quite distinct from the issue of attainment, or even from demonstrating reasonable further progress. By law, the agency had to ensure that all the transportation plans and programs were consistent with the air quality plans and programs in the SIP. For purposes of the conformity assessment, according to the MTC, it could rely on the initial assumptions used in making the attainment demonstration. It was not required to revise those assumptions in light of unexpected changes. In other words, the projects in the regional transportation

plan (RTP) and transportation improvement program (TIP) should conform as long as they accommodated the projected level of population and
development. Although pollution is clearly a function of the number of
automobiles on the road at a given time, from the agency's viewpoint, the
number of automobiles was dependent on population and land use, not the
number or location of highways. If the proposed highways served the
anticipated traffic, then their impact should be beneficial or at least neutral.
Only if a project increased congestion should it be considered potentially
harmful.

Although the court had rejected the MTC's qualitative assessment procedures and ordered the agency to develop a more quantitative approach, the
MTC took the position that this still only required it to show that constructing
the system would not interfere with the objectives of the Plan. In general,
that meant making sure that the planned highway system did not increase
congestion and so result in more vehicle emissions. It did not view the
procedure as requiring the MTC to redo its cleanup plan every time a new
freeway project was proposed. In other words, it did not require the agency
to prove that particular highway configurations would result in attaining the
clean air standards, only that the plans and programs would contribute to
reasonable progress in implementing the TCMs. Nor did it mean that each
individual highway project had to actually reduce air pollution as long as the
entire system when completed would contribute to improving air quality.

The Sierra Club, on the other hand, viewed the conformity assessment
procedures as an integral part of the overall cleanup strategy. Its position was
that the RFP standards in the Bay Area Plan had established a sort of
emissions "budget" that the MTC was obligated to meet and that the conformity assessment procedure should ensure that pollution from vehicles on
the roadways did not contribute to the region's being in nonattainment. The
plaintiffs believed that additional freeway construction would induce more
growth, and that in turn would produce more traffic and consequently more
pollution. They expected the MTC to show that each and every highway
project would contribute to pollution reductions consistent with the Plan or
at least would not aggravate the situation. If the MTC's initial travel forecasts
proved to be underestimated, then the environmentalists wanted the MTC to
either delay new construction until new control measures could be put in
place or redesign its highway proposals to compensate for the additional
pollution.

To comply with the court's order, the MTC proposed a new modeling effort that would provide the basis for determining whether the TIP and RTP amendments conformed with the Bay Area Plan. In many ways, the proposal represented an improvement over the standard practice, but the plaintiffs still found it to be unsatisfactory. The main dispute between the parties centered on the Sierra Club's contention that the MTC's travel model ignored the full effects from increased traffic. If conformity assessments are to ensure that new highway construction will contribute to achieving the clean air objectives, the models used to conduct those assessments need to present an accurate picture of how much pollution will be generated by the future transportation system. The Bay Area had been experiencing its share of new growth. The environmental groups in this case were very concerned that the MTC's proposed travel models did not reflect this increased growth and did not take sufficient account of the possibility that new highway construction would itself contribute to further growth and thus hinder efforts to improve air quality in the Basin.

The Sierra Club offered three main objections to the MTC's proposal. First, it complained that the way in which the MTC intended to carry out its assessments for transportation plans and programs would underestimate the actual amount of air pollution those facilities would generate. Second, the plaintiffs insisted that, in addition to plans and programs, each individual project that came up for approval should be subjected to a separate conformity assessment to see how much it would contribute to air pollution levels. Third, they felt that the proposal focused too much on the regional level and therefore could not adequately identify potential sources of CO pollution, which tend to occur as local rather than regional problems.

The Sierra Club charged that unless the assessment procedures were corrected, the MTC would continue to favor growth-inducing highway expansion projects over traffic reduction and would fail to require adequate mitigation measures. The MTC, on the other hand, believed that highway construction would help relieve congestion and improve average speeds, thus reducing excess emissions caused by stop-and-go traffic. From the agency's standpoint, any increases in overall travel levels would be compensated for by improvements in tailpipe emissions from tighter inspection and maintenance (I/M) programs and new car standards. The issues were complicated by adoption of the 1990 Amendments to the Clean Air Act, which the MTC contended rendered the provisions of the Bay Area Plan obsolete.

MTC's MTCFCAST Computer Model

The Bay Area Plan mandated conformity determinations for both the short-term TIP and the long-term RTP. Because the court's May 7th Order had interpreted conformity to require a quantitative air quality analysis, the MTC had to estimate future emissions from the planned transportation system. To accomplish this, the MTC proposed to create a computer model of the Bay Area transportation system for the year 2010, including all projects in the current TIP and implementation of all adopted TCMs, and to project both the number of daily trips and the average VMT associated with construction of that system. That information would then be combined with specific emissions factors to predict the amount of air pollution that would be generated from building versus not building the projects in the TIP.

The MTC's model development went through several iterations, both to respond to objections raised by the Sierra Club and to reflect changes required by the 1990 amendments. As described in Chapter 1, the standard travel demand modeling process, when adapted to estimating air pollution emissions, consists of a long line of stepwise models, including land use models, trip generation models, trip distribution models, traffic assignment procedures, pollution emission models, and pollutant dispersion models (see Figure 1.1). Developed for evaluating alternative shapes and sizes of regional transportation networks, these transportation planning models were simply not intended to be used for detailed evaluations of pollutant emissions. Figure 3.1 presents a diagrammatic view of the various steps in the MTC's proposed new emissions modeling process. Starting at the top of the page, the process begins by using subcounty population, housing, and employment growth projections from ABAG's POLIS model,[2] supplemented with census data and household surveys, to generate a 2010 baseline. This information is fed into MTC's traffic forecasting model, known as MTCFCAST, which predicts travel demand for transit facilities, highways, streets, and roads.

The standard model is essentially static and does not account for the possible effects of changes in travel times or trip numbers on individual travel-related behavior. In contrast, MTCFCAST was designed to take account of the effect of price and service level changes on travel decisions. The model divides the region into 700 discrete geographical zones. It makes projections for five travel decision factors using five submodels, shown inside the double-line box in the figure. The five submodels, in order, are

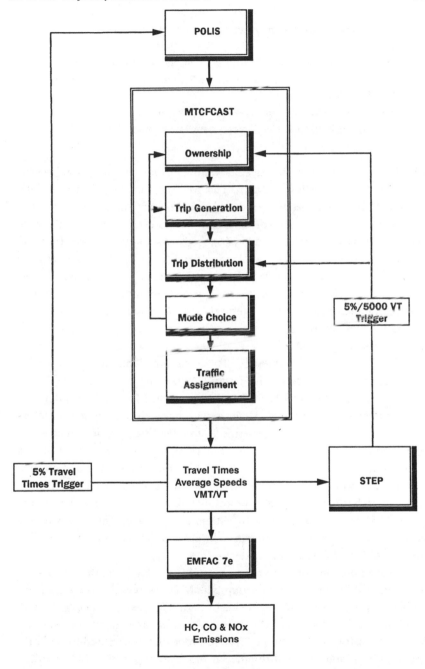

Figure 3.1. MTCFCAST Travel Forecast Model

1. auto ownership (whether members of a household will own 0, 1, or 2 or more cars),
2. trip generation (the number of trips per household per day),
3. trip distribution (the destination of each of the interzonal trips made by members of each household),
4. mode choice (whether to drive, walk, or take transit), and
5. traffic assignment (the selection of a transit or highway route).

Following the standard practice, the outputs from each submodel become inputs for the succeeding models. The MTC recognized, however, that additional highway capacity or reduced travel times from congestion relief could encourage higher travel demand and therefore offset some or all of the air quality benefits from higher average speeds. It sought to account for these indirect effects in its model. Two kinds of indirect effects were identified: (a) short-term *second-order* effects involving changes in individual or household travel decisions (trip generation, trip distribution, mode choice, and traffic assignment) and (b) longer-term *third-order* effects related to residential and business location decisions. Both sets of effects were incorporated into the modeling process by way of a series of feedback loops between several of the various submodels. The resulting dynamic model was a significant improvement over the earlier standard modeling approaches.

As illustrated in Figure 3.1, one loop connects the Mode Choice submodel to the Auto Ownership and Trip Generation submodels to account for factors such as the possibility that enhancing transit facilities might induce some trip takers to choose public transit instead of driving, or to choose not to own a car, or to own fewer cars. The modeling process is repeated until a reasonable equilibrium is achieved in the Mode Choice and Traffic Assignment submodels. The net result, represented by the shaded box below the MTCFCAST model, is a series of estimates for travel time and average speeds for the various road segments at different hours, along with total VMT and vehicle trips (VTs) for the system.

Because of the time and expense in rerunning the Trip Generation and Trip Distribution submodels to equilibrium, the MTC proposed adding a second feedback loop, shown on the right side of Figure 3.1, using a separate model known as STEP, which predicts the effects of changes in travel times on travel and destination decisions (second-order effects).[3] Should the STEP output show a significant difference in travel between the 2010 "build" condition and the result from modeling a hypothetical "no-build" alternative (consisting of the 2010 population and housing figures but the 1990 transportation network), MTC would incorporate the results back into MTCFCAST to determine the impact of these changes. If STEP displayed either (a) an increase of more than 5% in trip generation between districts that experienced at least 100,000 trips per day under "no-build" conditions or (b) an addition of 5,000 new trips between districts experiencing fewer than 100,000 trips per day, then the new estimates would be fed into both the Trip Distribution and Ownership submodels. Smaller changes would be considered statistically insignificant.

To handle the potential long-term impact of increased travel times on business and residential location decisions (third-order effects) that could affect future land use patterns in the region, the MTC proposed another feedback loop, shown on the left side of the figure. If total estimated travel times between the "build" and "no-build" alternatives exceeded a 5% "trigger" (the sensitivity threshold of POLIS to changes in travel times), the MTC would ask ABAG to rerun its demographic projections in the POLIS model, using the new computed travel times. The new results from POLIS would in turn be fed back into the MTCFCAST model. These loops account for how changes in the transportation network would affect trip patterns and even redistribute population and development in the region, but not the possibility that adding capacity could "attract additional growth from outside the Bay Area."[4]

Based on these calculations, the MTCFCAST model could be used to project the number of vehicle trips per day (and the zone of origin and destination), VMT in the region, travel times, and the average speed for each highway link for both peak and off-peak travel. This information would then be used to predict regional HC emissions and regional and subregional CO emissions from vehicular sources (represented by the shaded box at the bottom of Figure 3.1). This would be done by applying new emissions factors developed by the Air Resources Board (ARB), known as EMFAC 7e, which estimate vehicle emission from an "average" vehicle in the fleet traveling at various speeds. The conformity assessments would be carried out by comparing the impacts both with and without the proposed projects.

TIP Conformity Assessment

For the TIP conformity assessment, the MTC would run the MTCFCAST model to estimate the regional HC and CO emissions associated with the projected 2010 level of travel and would disaggregate its results by individual freeway corridors. A *corridor* is a geographic area in which the most significant travel effects of major projects would most probably occur. MTC would not perform conformity assessments for individual highway projects. Unless the corridor contained only a single "major project," in which case it would estimate emissions both with and without the project, but otherwise, all corridors with two or more major projects would be viewed as a unit.[5] The agency considered evaluating individual air quality impacts from multiple projects in a single corridor to be "meaningless" and simply refused to undertake such analyses.

MTC also proposed to project CO emissions within a 25 km^2 radius around CO hot spots associated with the 2010 "build" and "no-build" alternatives to satisfy the Areawide air quality standards. The proposed model, however, could not predict intersection-level emissions at individual hot spots, and therefore the MTC would rely on available traffic projections from local traffic authorities to estimate CO emissions at these locations.

RTP Conformity Assessment

For the RTP assessment, MTC proposed to follow a similar procedure to the TIP assessment, except that instead of merely modeling the "build" and "no-build" conditions, it decided, as part of its environmental impact review of the RTP amendments, to model four alternative scenarios containing different mixes of future long-term highway and transit system investment. On the basis of its analysis, the agency would select one of the following alternatives:

1. no project (no-build),
2. highway capacity (extending the current highway network to its foreseeable limit for 2010 with only minor transit improvements),
3. transit capacity (devoting all new facility funding resources to rail, bus, and ferry service and none to new highway projects), or
4. transit/highway blend (balance between highway and transit investment).

The 2010 horizon date for both the TIP and the RTP assessments was chosen to conform to ABAG's long-range projections of population and employment distribution for the Bay Area, and also to allow time for the projects in the TIP to be completed and for regional travel patterns to reach equilibrium.

MTC's Test for Conformity

As discussed in Chapter 1, at this time Department of Transportation (DOT) regulations defined conformity in terms of whether the projects in the RTP and TIP supported or interfered with the TCMs in the SIP, not whether the traffic they generated was consistent with the initial VMT or VT projections, nor whether the emissions from the resulting traffic system stayed within the baseline estimates. Therefore, the MTC's proposed test for conformity did not relate directly to the Bay Area Plan's pollution reduction targets for CO and ozone.[6] The agency's position was that the Contingency Plan was the primary mechanism to ensure that the region stayed "on the RFP line." If the TCMs did not produce sufficient reductions to meet the established emissions targets and as a result the region did not meet the federal air quality standards, this would trigger the Contingency Plan provisions and additional TCMs would have to be adopted to cover the shortfall. The MTC felt this to be its only obligation under the Contingency Plan and that RFP should be measured in terms of emissions reductions, not residual emissions levels. Because it did not believe it had to account for emissions due to excess growth in the region, beyond that initially projected in the Plan, from the agency's standpoint it could hardly be expected to measure highway plans and programs against the Plan's 1987 standards for residual pollution levels (430 tpd for ozone and 1,541 for CO), which depended entirely on the accuracy of the initial assumptions about ambient regional emissions levels.

In light of this reasoning, the MTC sought to dissociate the whole conformity question from the issue of NAAQS attainment or RFP. The MTC explained that TIPs and RTPs serve very different purposes from those served by a SIP: "There is no authority for the proposition that RTPs and TIPs should function as SIPs, or that conformity assessments—which are elements of RTPs and TIPs—should function as demonstrations of attainment."[7] The agency asserted that the proper scope of inquiry in assessing conformity with the SIP should be limited to whether the individual implementation measures, the TIP and RTP, were consistent with the original air quality plan, not whether the traffic associated with the system build-out would generate more or less pollution than allowed by the federal standards. Moreover, federal law did not equate the conformity assessment procedures with demonstrating attainment. In accordance with the 1977 Clean Air Act and DOT's guidelines, which placed their emphasis on employing TCMs for air pollution control (rather than comparing emissions level), the MTC agreed that the TIP or RTP amendments should support all the adopted TCMs.[8] Moreover, the agency also agreed with the plaintiffs that overall new highway projects should contribute to lower emissions. The agency contended that the correct test of conformity was therefore whether the plan or program

1. does not adversely affect TCMs in the SIP,
2. contributes to reasonable further progress in implementing the TCMs contained in the SIP, and whether
3. completion of the projects in the TIP or RTP would not entail an increase in emissions that would cause or contribute to a violation of NAAQS or delay timely attainment.

Based on its interpretation of federal law, however, the MTC argued that the Conformity Assessment in Appendix H did not require it to compare future levels of travel-related emissions to the RFP levels associated with *attainment of NAAQS* in the Bay Area Plan. Instead, conformity merely required that there be *no increase in emissions* from all the projects in the RTP or TIP. Moreover, it did not require any project-level reviews.

Unlike the RFP demonstration in the Contingency Plan (discussed in the next chapter), which must establish quantified emissions reductions that would meet the 1987 standards, the MTC asserted that neither the RTP nor the TIP had to be tested against those standards. The conformity assessment merely had to show the region was making progress toward compliance. Thus, the RTP/TIP would conform if projected vehicle emissions for the "build" condition in the horizon year were lower than under the "no-build" assumption, or at least lower than at present. In that case, there would be no increase in emissions that would "cause or contribute to an NAAQS violation or delay attainment." In other words, MTC could approve new highway projects as long as air quality would be better in the future if those projects were constructed, even if the area still exceeded the NAAQS.[9] Accordingly, to carry out the new test, the MTC decided to prepare a new baseline estimate of vehicle emissions beginning from the then-current year (the 1990 inventory) by extrapolating from its original data so it could make the appropriate comparisons.

The MTC defended its new conformity assessment procedures as meeting or exceeding the requirements in the Bay Area Plan and the 1977 Clean Air Act. The Sierra Club nevertheless charged that the procedure was defective and was more the product of agency judgments about "policy, cost-effectiveness and convenience, than the capabilities of traffic modeling."[10] Initially, the Sierra Club argued that the RTP/TIP assessments should demonstrate that emissions from the completed projects would be less than the Bay Area Plan's regional, Areawide, and Hot Spot "attainment level" inventories. Assuming the region had achieved NAAQS, or was at least maintaining RFP, the MTC's definition of conformity might not have posed any problems. Projects from a TIP or RTP that met the criteria—that is, which did not increase emissions—would not cause new violations or "delay timely attainment." What concerned the plaintiffs was that the Bay Area was not "on the RFP line," and indications were that pollution problems were getting worse. For example, the MTC's initial estimates for 1990 were considerably higher than the RFP targets, in part a reflection of the region's increased growth. This meant that projects that did not contribute to significant reductions in pollution, but allowed it to remain at these already high levels, would make it harder to bring the region into compliance and would place a greater burden on transportation and other control measures to compensate. For these reasons, the plaintiffs wanted the conformity assessments to prove that the RTP amendments and TIPs would meet the 1987 standards in the Plan.

The MTC continued to assert that the conformity assessments did not have to relate to RFP. Emissions levels from building a particular transportation network were not considered particularly relevant as long as constructing that system, accounting for any adopted TCMs, would not make pollution worse. (In that case, the plans and programs would neither "cause or contribute" to a violation of the ambient standards, nor "delay their attainment.") The agency held to its view that, in general, the configuration of the highway network was not a key determinant of regional traffic volumes or pollution levels. Highway construction would, if anything, be neutral or have a net beneficial impact on air pollution by reducing congestion. Delaying or even stopping highway construction would have little, if any, positive effect on regional air quality and could even be detrimental. Their strategy again reflected the prevailing reliance on TCMs and other mitigation measures, rather than making changes to the transportation system itself, to reduce pollution.

The Sierra Club, in contrast, believed that the conformity assessments should contribute to achieving the Bay Area Plan's attainment goals as soon as possible. Although the environmentalists eventually, if somewhat grudgingly, accepted the premise that the modeled RTP/TIP emissions levels need not be compared directly to the RFP targets—that is, conformity could be shown by comparing emissions for the "build" scenario (a) with the "no-build" case at some future year, and (b) with current (1990) emission levels—they nevertheless insisted that the MTC's proposal still failed to satisfy the requirements of the Clean Air Act.[11] The plaintiffs felt the MTC's review was systematically biased in favor of highway expansion—that it was designed to approve any projects where the agency could show at least some pollution reduction for the "build" alternative, regardless of whether federal air quality standards were violated or even whether the project interfered with required TCMs. For example, they suggested that reducing congestion could actually undermine efforts to encourage transit ridership.

The MTC insisted, however, that its modeling approach was state-of-the-art and that plaintiffs had no valid objections but only wanted the agency to use a different modeling technique more favorable to their political agenda. The MTC contended that its new methodology was "virtually unique" among MPOs and that it was akin to the advanced sort of modeling then being debated in Washington as part of the amendments to the Clean Air Act.[12] No other MPO was using computer modeling in conformity assessments in the manner being proposed by the MTC.

Deficiencies in the Modeling Process

Beyond disputing the MTC's view of what constituted being in conformity, the Sierra Club also complained about the nature of the modeling process itself and how it would be applied. However conformity was defined, the proposed conformity assessments would be valid only insofar as they presented an accurate comparison of the potential consequences between building and not building the projects under evaluation. The plaintiffs' response to the MTC's new procedure centered on three main objections. First, the Sierra Club contended that the plan and program assessments were inadequate and biased toward approving new projects. They felt that the MTC did not take account of increased growth in traffic, especially that resulting from highway expansion. Plaintiffs also objected that the agency proposed to use methodologies and assumptions that were inconsistent with those used for making the original projections in the Bay Area Plan, rendering any conformity findings meaningless.

Second, they insisted that the MTC should subject each project in the TIP to a separate, quantitative conformity assessment before approving it, rather than simply assessing all projects in the TIP as a whole. By focusing on establishing conformity at the plan and program level, the plaintiffs believed MTC was ignoring local air pollution effects caused by individual highway projects. Unless individual highway projects were subjected to conformity assessments, some projects might be approved even though they made air pollution worse. Finally, they worried that because the proposal ignored subregional effects, it would not be capable of identifying potential sources of localized CO pollution (hot spots). According to plaintiffs, in accepting the projects' sponsors claims without any independent confirmation, the MTC was violating its obligation, as the designated MPO, for ensuring conformity. To summarize, the plaintiffs' basic position was that the proposed modeling effort did not accurately reflect all the potential pollution effects of the MTC's transportation plans, programs, and projects and therefore could help to justify highway construction that would in fact have significant negative environmental consequences. As a result, the Bay Area would not be able to achieve the NAAQS as required in the Clean Air Act and the Bay Area Plan, at least not without even more extensive and costly control measures.

The issues raised here by the plaintiffs involved both legal and technical considerations. Transportation planning methods are complex and multifaceted. They require the collection of specialized data sets, which are not used by other planners or for other purposes, and they require mastery of survey and sampling methods as well as several independent modeling procedures, which are used in conjunction with one another to forecast the future use of transportation facilities under different conditions. Given the arcane nature of these methods and techniques, it is not surprising that the parties to the suit enlisted the help of technical experts in this field.

The plaintiffs called on Dr. Peter Stopher, a nationally recognized expert in transportation planning methods, whose career has included professorships at several leading universities and service with well-known transportation consulting firms. Dr. Stopher, who holds a doctorate in transportation planning methods from Northwestern University, taught these methods at Northwestern and Cornell universities and presently is a Professor of Civil and Environmental Engineering at Louisiana State University in Baton Rouge. He has also been a consultant to many transportation planning programs, including those in Miami and Los Angeles. Dr. Stopher is the author of a well-known textbook on transportation planning methods and is widely recognized as an expert on the application of survey research and sampling methods to transportation planning.

The MTC retained the services of Dr. Greig Harvey, vice president of a small, highly regarded transportation planning firm in Berkeley, California. His areas of expertise include travel behavior theory, travel demand analysis, and the organizational and institutional setting of transportation. Educated at MIT and the University of California, Berkeley, Dr. Harvey was for a time an Assistant Professor of Civil Engineering at Stanford. He has helped a number of metropolitan areas modernize and improve their transportation forecasting methods, and recently prepared a manual on transportation and air quality analysis that was published by the National Association of Regional Councils.

Confronted with highly technical subject matter related to data collection, modeling software, disagreements over the strengths and weaknesses of travel demand forecasting models, and contradictory claims of the models' capabilities and limitations, the court decided that it also needed technical assistance to understand and incorporate into its rulings some knowledge about the methods and information sources at issue in this case. Although the appointment of a neutral expert is rare, even in civil cases, the court has inherent authority to do so. The plaintiffs and the defendants each submitted to the court the names of two technical experts whom they considered unbiased and appropriately knowledgeable, and who had no prior involvement in this case. The four nominees were asked to submit resumes listing their qualifications and background and to say whether they would be willing to serve as a court-appointed expert. Judge Henderson reviewed the qualifications of the nominees, conferred with lawyers representing both plaintiffs and defendants, and selected one of the authors, Dr. Martin Wachs, Professor of Urban Planning at UCLA, where he is presently the Director of the Institute of Transportation Studies. Dr. Wachs was a doctoral classmate of Peter Stopher at Northwestern University and is considered an expert in transportation planning and policy. At the time of his appointment as neutral expert, he was conducting research on the relationships between transportation investments, land use patterns, and air quality. After accepting the appointment, Dr. Wachs attended all major court hearings in this case, read technical documents prepared by all parties to the litigation, and on several occasions met with lawyers representing both sides to seek clarification of their arguments. He not only joined the judge and his clerk as they crafted the wording of the court's rulings but also read several drafts of rulings before they were released to ensure their technical accuracy with respect to transportation planning methods and procedures.

To better understand the decisions ultimately reached by the court, we will next take a closer look at the plaintiffs' main objections to the MTC's initial conformity assessment proposal, namely (a) the adequacy of the MTC's method for determining conformity for transportation plans and programs, (b) the lack of project-level conformity analyses, and (c) the failure to monitor CO hot spots. The passage of the 1990 amendments led the MTC to make several revisions to its initial modeling approach and forced the court to consider the impact of the amendments on the agency's revised proposal. That issue is taken up at the end of this chapter.

Determining Plan and Program Conformity

The parties vigorously disputed how conformity assessments should be made with regard to future transportation plans and programs. The Sierra Club pointed out a number of ways in which they believed the plan and program conformity assessments would systematically underestimate pollution from new highway construction. First, the plaintiffs objected to the length of time the MTC had given itself in which to demonstrate conformity. They complained that the proposal assessed only long-range conformity, not short-term compliance. Because long-range estimates are less reliable, highway projects could increase pollution in the short run, with little assurance that conditions would in fact improve over time. Second, the plaintiffs challenged the MTC's use of new emissions factors to estimate pollution levels instead of those it had used to develop the RFP standards in the Bay Area Plan. They accused the MTC of making assumptions about average vehicle emissions in future years based on untested and untried tailpipe standards, which tended to underestimate probable vehicle emissions and therefore minimized the potential impacts of increased traffic. Third, plaintiffs asserted that by using these optimistic forecasts to justify new construction, the MTC would only aggravate the problem, because added highway capacity would encourage even more growth and development than first projected and in turn add to the region's pollution problems. In their view, the model did not do a sufficient job of assessing the impact of new highway construction on future growth in regional traffic levels. Because of these deficiencies, plaintiffs argued that the plan and program conformity reviews would permit increased traffic and more pollution. The MTC believed equally strongly that expanding highways would reduce congestion and thereby improve air quality. Each of the three key issues raised by plaintiffs—(a) the long time horizon, (b) the choice of appropriate emissions factors, and (c) the impact of increased highway capacity—is described in greater detail in the following.

Long Time Horizon

The Sierra Club first objected to the MTC's proposed time frame for determining conformity. As noted earlier, MTC proposed to evaluate the projected emissions from the RTP and TIP in the horizon year 2010. Clearly, because the statutory deadline for NAAQS had already passed, the MTC had to choose some future date for assessing compliance, reflecting the construction time of the projects in the plans being evaluated. Plaintiffs charged, however, that by using the year 2010 for projections of air quality, rather than some earlier date, the MTC was simply abandoning efforts to achieve air quality improvements in the short term. Pollution could remain unacceptably high, or even increase over the next decade or so, provided the model showed improvement later on. By choosing a 2010 horizon date, the MTC would be able to approve projects that might do little to improve air quality for years. By the time it would become obvious that the NAAQS could not be achieved, many projects would already have been constructed or would be too far along to stop. As a result, the MTC gained a long extension of time for itself to demonstrate RFP, which was not authorized by the Clean Air Act. Because there was no assurance that the region would meet the pollution standards anytime before 2010, in their view the whole procedure violated the Clean Air Act's mandate to attain the air quality standards "as expeditiously as practicable."

The MTC responded that because *conformity* was not to be equated with *attainment,* the choice of a horizon date was really irrelevant. Besides, the proposed time frame was reasonable, because it was not practical to do projections every year and the 2010 horizon year coincided with the long-range planning horizon for the region, as developed by ABAG. Interim-year models would not show as great an impact from increased capacity and congestion relief. Furthermore, a long time frame was necessary to evaluate the network as a whole inasmuch as highway and transit investments may have impacts on land use, travel, and emissions that continue to build over many years.[13] The agency had therefore decided to model conformity in 2010 so that any apparent short-term air quality benefits did not disguise possible long-term detriments from expanded roadway capacity. MTC also claimed that near-term modeling would require it to speculate which projects in the TIP would be completed by a particular year, whereas the long-term approach allowed for full build out and consideration of their synergistic effects.

The Sierra Club replied that because Section 176(c) of the Clean Air Act required conformity without reference to time, MTC could not avoid short-term environmental effects by adopting only one horizon year. They also challenged MTC's ability to know in advance the relative effects of short-term versus long-term modeling, so it would be arbitrary to choose one approach over another.[14] Plaintiffs urged the court to require that the MTC demonstrate conformity continuously over at least a 20-year period (using appropriate interim dates) even if that meant revising its project approval process to attain more certainty in construction schedules. Thus, the plaintiffs concluded, the MTC's highway construction plans should be evaluated to ensure sufficient emission reductions for "1991, 1995, 2000 and beyond," not just by 2010.[15]

Choice of Emissions Factors and VMT

The Sierra Club also objected to the MTC's use of updated mobile source emissions factors, in its travel demand model, to estimate future pollution levels; they insisted that the agency had to base its calculations on the same methods used to support the control strategies in the Bay Area Plan. They pointed out that the new EMFAC 7e factors assumed that the regional vehicle fleet would be far less polluting, due to anticipated improvements in engine technology, than the EMFAC 6c factors employed in preparing the SIP. Because this new set of factors incorporated California's tougher new car tailpipe regulations, the projections would generate even "cleaner" emissions estimates than using the previous factors, which did not reflect these revised standards. As a result, the estimates of future pollution associated with full build-out would be lower than those derived using the original EMFAC 6c factors. The Sierra Club took the position that the comparison between present emissions estimates and those from the "build" and "no-build" scenarios should be consistent with the assumptions and methodologies used in the Bay Area Plan.

Plaintiffs also urged that using the new tailpipe standards to evaluate the impact of added highway capacity would have the effect of permitting even higher traffic levels in the future. This was because the results obtained using the new emissions factors would overstate the air quality benefits of new freeway construction. By modeling a hypothetically cleaner fleet, the MTC would be able to claim immediate air quality benefits from standards that would not even take effect before the end of the century. On the basis of these unproven measures, the MTC could claim that although there would be more cars on the road, the amount of pollution per vehicle would be less than initially assumed. Thus, the region could exceed the Plan's projected number of vehicle trips and VMT and still the MTC could claim it was making RFP. The Sierra Club argued that the EMFAC 6c factors were "enforceable control measures" and urged the court to require the MTC to use these same emission factors to maintain consistency. The plaintiffs contended that using the new factors to justify compliance, rather than the initial assumptions, amounted to modifying the attainment demonstration in the Plan without going through the formal SIP revision process.[16] Echoing an argument first made by the MTC against the plaintiffs, the Sierra Club maintained that the agency should not be allowed to unilaterally change its approach for achieving clean air but must carry out the strategies originally adopted in the Plan and approved by the EPA.

The MTC insisted, though, that it was entitled to use the current best available estimates for growth, development, travel, and vehicle technology, including ABAG's latest population and employment forecasts, the MTC's MTCFCAST model for future travel demand, and ARB's current emissions factors. These did not constitute new control measures or changes to the SIP. Moreover, the MTC argued, the Bay Area Plan did not commit it to achieving any particular levels of VMT or VT.

Putting the issue in a broader context, the MTC addressed what it believed the basic question in a conformity assessment should be, What will the impacts be on air quality from adopting a plan that includes a given number and type of highway, transit, safety, repair, and other projects? The agency objected that the plaintiffs wanted it to redo its models, using the same assumptions about tailpipe emissions as in the 1982 Plan, even though *actual* trip volumes would be higher and emissions lower. Because current VMT and trip levels would have to be applied against "fixed and now outdated emissions factors," virtually a single additional trip or mile traveled would "throw the TIP out of conformity." The MTC feared that unless it could employ the new information, it would be prevented from approving a balanced RTP or TIP that included not only "congestion-relieving highway projects" but also "heavy investments in mass transit."[17] It was precisely those future "congestion-relieving" projects that most concerned the lawyers for the Sierra Club.

Impact of Improved Capacity on Regional Growth

Beyond the questions of horizon years and appropriate emissions factors, the Sierra Club raised a more fundamental objection to the MTC model. It argued that although the MTCFCAST, STEP, and POLIS models captured some of the impact of increased highway capacity on traffic levels, they did not fully consider either the extent of behavior changes in response to new highway capacity or the effect improved capacity might have on the level of regional development.[18] In effect, the Sierra Club suggested, by not considering the relationship between freeway expansion and growth in travel demand, the models would simultaneously underestimate the traffic impacts from implementing the "build" scenarios and overestimate those from following the "no-build" alternative.

Plaintiffs firmly believed that the MTC was building highways on the assumption that vehicle use would increase, instead of attempting to reduce traffic. The MTC's proposed model took as given that the region would continue to grow regardless of whether roadways were congested. They worried that building additional highways to relieve congestion would simply encourage more driving and result in more pollution. According to the plaintiffs, the transportation system in the Bay Area was already so congested that it served to dampen new development, but that constructing new facilities with additional capacity would merely increase travel demand, producing less of a reduction in congestion than expected. New highway capacity could also lead to increased levels of development as well as a different regional distribution of housing and jobs. The MTCFCAST model, though, looked only at whether added capacity would make traffic run smoother. Thus, it overstated the ability of new highways to relieve congestion and understated overall levels of future vehicle use.

According to the plaintiffs' transportation expert, Dr. Stopher, a capacity-constrained system, like that in the Bay Area, has a pool of unsatisfied or "latent" travel demand. That is, many highway and transit facilities are overloaded for substantial periods of the day, resulting in low average travel speeds and prolonged peak periods. As a result, travelers may forego trips, leave work earlier or later, find alternate routes (leading to congested arterial streets), "chain" trips together, choose closer destinations, or change their travel modes. Eventually, they may even relocate their residences.[19]

Because most studies show that average travel time has remained relatively constant over the years, adding additional lanes or new connections to an existing highway system already operating at capacity, in an attempt to reduce congestion, could result in increased vehicle use as drivers shift their commute trips back toward traditional working hours, make trips they otherwise might not take, drive solo, or switch from public transit to private automobiles. In addition, added system capacity might encourage an increase in regional development or facilitate development in more distant areas, minimizing the benefits from reducing traffic congestion. Most transportation models, however, do not adjust the projected number of vehicle trips based on the supply of transportation facilities.

In the MTCFCAST model, for instance, the estimated number of trips from the Trip Generation model is initially assigned to traffic corridors under assumed "free-flow" conditions, which may result in some links operating over capacity. Speeds are then recalculated and trips are reassigned to other links, using a "capacity-restraint" function. The results are fed back into the model where they affect trip destinations and mode choices, but, as explained by Dr. Stopher, "congestion [has] no effect in the model on the distribution of population and employment forecasts, nor [on] the total forecast of trip-making by the population in the region."[20] In other words, no change is made in the projected amount of regional travel despite changes in average speeds or travel times.

With more capacity resulting in higher speeds, plaintiffs argued that destination and route choices would be affected, leading to longer trips, more frequent trips, or both. If these effects were not incorporated into the assessment process, it was possible that the "build" analysis would overestimate the benefits of capacity increasing projects and simultaneously underestimate the negative impacts of these projects. Similarly, a realistic assessment of the "no-build" scenario would need to consider how increased congestion might divert travel away from morning and evening peak hours, eliminate some trips, or encourage alternate travel modes, and also how it might affect the total level and distribution of new residential and commercial development.[21]

Because the model focused on the year 2010, without regard to whether RFP was achieved and maintained in the meantime, plaintiffs felt it ignored short-term effects caused by less congestion drawing out increased traffic. Although the MTCFCAST model is capable of assessing, through its feedback loops, the impact of capacity changes on destination, route, and mode choices, it cannot account for their effect on changes in the duration of peak periods, numbers of trips either made or foregone, or the practice of trip "chaining." More important, according to Dr. Stopher, it could not predict changes in land use and development in the region.[22]

The MTC countered that even if there was a link between highways and development, there was no practical way to model the connection. According to the MTC, there was no proof that the overall levels of regional development might be affected by highway construction. Rather, regional development is driven by economic and other factors outside the scope of transportation planning. The agency believed its model did the best job possible of predicting behavior changes from increased capacity. The issue was not whether to account for changes in travel behavior or location decisions but *how,* and this was a matter within the MTC's administrative discretion.[23]

The agency also insisted that the system was not "capacity-constrained," as plaintiffs charged—any so-called new traffic was just a shift from other routes, times, destinations, or a combination of all three. Its POLIS model took account of changes in travel times and captured changes in the distribution of development, though not the effect on overall regional growth, and the STEP model took account of changes in trip generation and trip distribution. Although it was true that MTCFCAST did not predict the effects of increasing highway capacity on the level of development throughout the region, the MTC pointed out that there were no models available that could predict such changes, as opposed to shifts in the pattern of development.[24] In short, the MTC saw no reason to account for any additional growth that might be attracted to the region merely because of its transit investments or, conversely, for the possibility that some persons might choose not to move to the Bay Area solely because some planned transportation improvements might not take place.

Lack of Project-Level Reviews

The second major dispute was over whether each individual project in the TIP would have to undergo a separate conformity assessment or whether the MTC could approve individual highway projects that might increase existing pollution merely because they would be part of a larger conforming plan or program. Specifically, the environmentalists' lawyers objected that the MTC's refusal to perform project-level conformity analyses conflicted with Appendix H of the Bay Area Plan, which expressly called for "an assessment of major highway projects to determine if they will adversely affect emissions" (see Appendix A, this volume). This provision, they asserted, carried out the intent of Section 176(c) of the Clean Air Act prohibiting approval of any "project, program or plan" that did not conform to the applicable SIP. Without individual assessments, the RTP and TIP could contain projects with significant negative impacts as long as they were balanced out by other projects with positive impacts on congestion and air quality. The potentially harmful projects could also be the first constructed, without any guarantees that the region would later receive compensating benefits from those other projects or even that they would ever be built.

The Sierra Club was particularly concerned that the proposed conformity assessment did not cover highway projects previously approved under the invalidated "qualitative" approach but not yet under construction. Also, it complained that despite the court's May 7 Order, the MTC had been continuing to grant final approval to new projects without subjecting them to any further conformity evaluation, instead of suspending action until the court could approve the new procedures. Plaintiffs contended that this was plainly inconsistent with Section 176(c)'s mandate that conformity determinations precede project approval. Their objections to this practice would become the subject of a separate injunction motion, which is discussed later.

Not surprisingly, the MTC took the position that neither the Clean Air Act nor the Bay Area Plan required project-by-project conformity assessments. To model regional air quality with and without each potential highway project in the TIPs and RTP amendments, as the plaintiffs were suggesting, would clearly be an enormous undertaking. The agency maintained that the language in Appendix H calling for an "assessment of major highway projects" had to be read in the context of performing the overall TIP evaluations. The MTC's lawyers asserted, moreover, that the court's May 7 Order did not require project-level emissions evaluations. The agency's position, which is now reflected in the 1990 amendments, was that an individual project conformed with the SIP as long as it came from a TIP that on the whole conforms with the SIP. The MTC's consultant, Dr. Harvey, defended the practice of evaluating all the projects in the TIP together, rather than making individual project-level assessments of emissions, in the following way:

> It is my opinion that the most meaningful overall approach to predicting the air quality impacts of the TIP, of major projects in the TIP, and of the RTP amendments, is to estimate the emissions which would be experienced from construction of the projects in the TIP or the RTP, and then to assess the contribution of projects in specific corridors to the total emissions projected. I agree with MTC that it would not be a sufficient exercise, for the purpose of assessing air quality impacts, to attempt to estimate quantitatively the emissions associated with a particular project by attempting to compare the difference, if any, in emissions associated with the "build" and "no-build" decisions, in isolation from other major projects in a corridor. The reason is that such an approach ignores the synergism of the particular project with other projects in a corridor.[25]

The MTC contended that it would be nearly impossible to estimate emissions levels for a particular project X in relation to all other possible combinations of projects from the TIP or RTP. As a practical matter, in some cases projects in a corridor may intersect with one another so that considering each one of them in isolation would ignore their combined effects. In other cases, projects may run parallel and therefore serve the same group of trip takers. According to MTC, these must also be considered together or their overall emissions impacts would tend to be understated. Segregating projects from one another in a single corridor would simply not produce a meaningful analysis of their true impacts on travel and air quality.

Furthermore, the MTC argued that two previous interagency agreement, discussed earlier, between EPA and the Federal Highway Administration (FHWA), which must authorize all projects constructed with federal funds, had ruled out project-level conformity reviews.[26] The approach in that joint agreement, which was later incorporated into FHWA regulations on conformity in federally funded transit programs, interpreted project conformity in the context of implementing TCMs, rather than whether emission targets were met and did not require any air quality analyses. On the basis of those regulations, the proper test was simply whether

1. the project is a TCM from an approved SIP, or
2. it comes from a conforming TIP, or
3. it is a project, exempt from TIP requirements, which does not adversely affect the TCMs in the approved SIP.[27]

In essence, under the DOT regulations, specific projects in the TIP would conform if they came from plans that conform, even if the individual project in question might adversely affect air quality or result in violations of the clean air standards.[28]

The reasoning behind the MTC's position here is that transportation plans and programs contain many projects, some of which may improve air quality whereas others may be detrimental but are needed for the smooth operation of the entire system. The key issue for purposes of conformity is whether as a whole the proposed projects have a net positive impact. Therefore, until EPA requires a SIP revision, an individual project conforms simply by being included in a conforming TIP. Most projects included in the TIP and the RTP are in a preliminary stage of development anyway, so their potential impact is hard to evaluate. Once they are refined, they must be subjected to a complete environmental review, which, the MTC insisted, would "also serve to ensure conformity" (see Appendix A, this volume).

The Sierra Club nevertheless continued to object that the MTC's proposal to combine individual projects into corridors for the conformity analysis violated the Clean Air Act. It argued that Appendix H independently required the MTC to conduct an individual review of all "major highway projects" and that this meant each project in the TIP must conform to the SIP in addition to overall TIP conformity, because Section 176(c) applied to "projects" as well as "programs" and "plans." This, it maintained, required the agency to make a specific finding that each proposed project neither "contributes to a violation of the ambient standards . . . nor delays their attainment."[29]

Plaintiffs insisted that the MTC's reliance on FHWA regulations, to limit its responsibility for making conformity determinations to plans and programs, was misplaced. The FHWA regulations governed only procedures for TIP certification, they argued, and just reflected the minimum requirements for federal funding—they were not a limitation on the control measures in local SIPs. According to the Sierra Club, the Bay Area Plan was broader in scope, because it was actually reviewed and approved under criteria in the EPA's 1981 SIP approval policy, not the FHWA regulations, and therefore had contained stricter standards in order to comply with the provisions of the 1977 Clean Air Act prohibiting MPOs from approving any nonconforming projects.[30] Those criteria were in turn based on the EPA's own interpretation of Section 176(c) contained in an Advance Notice of Proposed Rulemaking, which had concluded that proposed federal actions "should not be allowed to cause a delay in the attainment or maintenance of the [NAAQS] in any state."[31] According to the plaintiffs, Appendix H carried out the intent of the Advance Notice and the final SIP approval policy by requiring separate conformity assessments for "all major highway projects" in the TIP. These project assessments, plaintiffs concluded, like those for the RTP and TIP, should be determined by an independent "evaluation of emissions" rather than by reference to the criteria in the FHWA regulations, and without reliance on uncertain future findings and mitigation measures that might appear in the final environmental documents. MTC countered that the Advance Notice urged use of "existing review procedures" and certainly did not mandate any particular procedures for assessing conformity. Moreover, the agency argued that in approving the 1980 guidance document and subsequent FHWA regulations, the EPA had implicitly abandoned any of its preliminary ideas regarding conformity in favor of not requiring emissions evaluations for highway projects.

CO Hot Spot Analyses

Finally, the Sierra Club faulted MTC for not conducting any independent CO hot spot analyses but instead merely accepting the project sponsors' data. There was no plan to monitor projects for potential CO violation. Without independent confirmation, plaintiffs alleged, the agency could not meet its obligation to demonstrate RFP for CO emissions on a subregional basis. The problem was compounded by the MTC's refusal to conduct project-level conformity assessments, as discussed earlier. Because CO pollution is primarily a local phenomenon, the lack of project-by-project conformity reviews makes it harder to identify potential problems.

Specifically, the plaintiffs objected that the MTC's proposal did not require traffic forecasts for the CO Hot Spot at First and Santa Clara, in downtown San Jose. MTC claimed that its MTCFCAST model was simply not precise enough to do intersection-level analyses, and that because there were no other models available that were capable of doing them, it was reasonable for it to rely on the project sponsors' own data. In any event, as stated in Appendix H, all individual projects must undergo an environmental review by the MTC, which, according to the agency, would also ensure each project's conformity to the SIP.

The Sierra Club nevertheless insisted that the MTC should prepare an independent forecast of hot spots using any available means, or at least independently verify the sponsors' assumptions and methodology, so as to demonstrate consistency at both the subregional and census tract levels. It argued that MTC could not abrogate its responsibility for this just because it would be "costly" or "time consuming."

Plaintiffs concluded that the MTC's revised conformity procedures fell short and that the agency's interpretation of its obligations was contrary to the "overriding purpose" of the Clean Air Act "to protect the public from air pollution above all else."[32] To summarize, the Sierra Club objected to how the MTC proposed to carry out the conformity assessments in that (a) MTC used favorable assumptions of engine emissions to reduce the impact of otherwise nonconforming highway projects and did not guarantee real emissions reductions in the short term; (b) it failed to account for the impact of added capacity on regional growth and therefore underestimated future pollution levels; (c) it improperly avoided doing any project-level reviews; and (d) the impacts on air quality at potential hot spots were not independently evaluated. Moreover, although the individual submodels had been previously tested, the MTC had never evaluated the entire model in the way it presently intended to use it. The MTC protested that the plaintiffs were demanding a procedure that would entail "endless modeling, lengthy rounds of pointless 'validation,' " and years of waiting for modeling technology to catch up with human behavior.[33] It saw its own proposal as both meaningful and realistic by contrast.

As the court struggled with the competing contentions over the conformity assessment provisions, the plaintiffs sought to block the defendants from proceeding with several previously approved highway projects. The court was forced to address whether highway improvements in the Bay Area should go on while the conformity assessment procedures were being finalized. This issue brought to the fore the impact of the recently enacted 1990 amendments on existing transportation planning practices.

Injunction Against Highway Projects

Despite Judge Henderson's May 7 ruling that the MTC's conformity assessment procedures violated the Clean Air Act, the MTC had continued to process highway projects in the 1990–94 TIP even as it worked to revise those procedures. The MTC based its approvals on its own Resolution No. 2107 (adopted October 30, 1989), which was designed to carry out the Contingency Plan provisions of the Bay Area Plan for delaying any highway projects found to have an adverse impact on air quality. Each project in question had been subjected to the evaluation called for in the resolution. Because the MTC had already determined that these projects would not have a significant effect on air quality and therefore did not need to be delayed, the agency maintained that it could proceed with construction.

The Sierra Club sought a permanent injunction to set aside the approvals and to prohibit the MTC from approving any further projects that would create additional highway capacity until they could be tested under new, valid conformity procedures. The Sierra Club specifically objected to four major "highway expansion" projects worth close to $80 million,[34] asserting that the Clean Air Act granted the district court authority to enjoin construction projects approved in violation of SIP requirements, in order to prevent irreparable environmental injury.[35] The injunction motion brought home to Bay Area politicians and residents just how serious the relationship between transportation planning and air quality had become. The prospect of a halt to freeway construction drew an immediate public response and received wide coverage in local newspapers. It also had a direct impact on the litigation itself. A number of elected Bay Area officials and the City of Berkeley requested and received permission to file amicus curiae (friend of the court) briefs in support of the Sierra Club's motion. They asserted that permitting the MTC to approve nonconforming projects would only result in greater costs later on to mitigate the negative effects. On the other side, several cities, including San Jose and Mountain View, which were also affected by the disputed projects, joined with Santa Clara County and the projects' sponsors, the Alameda County Transportation Commission and the Contra Costa Transportation Commission, to support the MTC's position. They filed papers with the court, urging it not to delay what, in their view, were much needed congestion-relieving projects.

The plaintiffs and their supporters argued that the MTC was violating both the court's prior order and the agency's independent obligation, under Appendix H and Section 176(c) of the Clean Air Act, to determine conformity *before* granting final approval to any projects. They did not consider the procedures in Resolution 2107 to be adequate substitutes for the required conformity assessment inasmuch as the procedures were based on the same "qualitative" process that the court had already rejected. Under Resolution No. 2107, the only projects that would be listed for possible delay were those where the completed final environmental impact report (EIR) showed significant adverse environmental impacts and lacked required mitigation measures. All others would be allowed to proceed.

The Sierra Club objected to the MTC's relying on the environmental review process to justify proceeding with these projects. The EIRs would merely assess whether the projects posed a significant environmental hazard, not whether they were consistent with the Bay Area Plan. The mitigation measures listed in the EIR could be used to excuse increases in emissions and support making a finding of no adverse impact on air quality. The procedures outlined in the resolution also failed to consider whether proposed projects could undercut emission controls in the SIP or any adopted contingency measures. For instance, adding capacity could attract drivers from high-occupancy vehicle (HOV) lanes or public transit. In addition, plaintiffs contended, EIRs generally compare only the differences between building or not building the particular projects, differences that are often slight. Finally, there was no analysis of the highway projects' effects on VMT or daily trip numbers, nor did they evaluate the extent to which adding capacity may affect travel demand or regional growth. In short, plaintiffs insisted that the procedures for delaying individual projects due to the failure to meet RFP should not be used to make conformity determinations because they would overstate the air quality benefits of future planned highway projects.

MTC countered that the May 7 Order only invalidated the *manner* of determining conformity but did not specifically address any prior conformity determinations that had been made, and that therefore it could continue to rely on those previous findings.[36] The court had never ruled that the 1990–94 TIP, of which the disputed projects were part, did not conform to the Bay Area Plan. Moreover, the MTC argued that the FHWA, the agency required by law to ensure conformity, had certified the MTC's procedures, and the court should defer to its administrative judgment.[37] Plaintiffs had no hard evidence that any of the projects would harm the environment in any way and therefore, in the absence of a "substantial risk" to the public health, an injunction would be inappropriate.

The MTC also noted that none of the four voter-approved projects involved any new highway construction, only widening and upgrading, and that both the appropriate environmental reviews and its initial modeling for the revised conformity assessment showed that they would have beneficial effects on air quality. Mostly, though, they objected that any delays would increase the costs of all currently scheduled projects by as much as $1 million per month. Plaintiffs insisted, however, that in light of the Court's May 7 ruling, the burden was on the MTC to show that each project conformed with federal standards.

The district court considered both the injunction motion and the plaintiffs' objections to the proposed conformity assessment procedures on November 14, 1990. As a preliminary matter, the court noted that the rhetoric between the parties had escalated:

> Each side paints the other as trying to impose its own political and environmental agenda upon the public. Plaintiffs assert that "freeway happy" MTC continues to ignore the mandate of the Clean Air Act in its relentless pursuit of highway expansion as the primary solution to problems of congestion and pollution. According to MTC, however, it is plaintiffs who are improperly trying to impede projects that they view as "inconsistent with their [own] vision of ideal future urban land use patterns, population limits and densities, and other political and social issues."[38]

Judge Henderson reminded the parties that it was not the court's role to resolve complex and controversial social issues, but that its sole function was to enforce compliance with the strategies adopted in the Bay Area Plan.

The judge then held that inasmuch as he had previously determined that the original assessment procedures were "wholly inadequate" to meet federal standards, the MTC could not rely on any conformity determinations made under those disapproved standards. The conformity provisions in the Plan were designed to satisfy Section 176(c)'s prohibition on approving any nonconforming plans or programs. The "plain language" of the Conformity Assessment, the court concluded, imposed an affirmative obligation on the MTC to ensure that projects in the TIP conformed before giving final approval. By continuing to approve projects that had never been validly shown to conform, the court declared, MTC "only exacerbates the violation" the court found in its May 7 Order. The court also rejected MTC's argument that its obligation to determine compliance was met because the FHWA had determined that the 1989–93 TIP conformed with the SIP, ruling that although the FHWA may be the "ultimate arbiter" of conformity with respect to federal funding, the agency's actions had "no bearing on MTC's independent obligation to affirmatively demonstrate conformity under the Bay Area Plan."[39]

Finding that the MTC could not rely on its prior conformity determinations, though, did not end the inquiry. By then, the recently enacted 1990 amendments to the Clean Air Act had modified the rules relating to conformity determinations. At issue was whether the so-called grandfather provisions of the amendments overrode MTC's SIP obligations to make new conformity determinations prior to giving final approval to the four highway projects. On the basis of its review of the legislation, the court concluded that they did not.

Effect of the 1990 Amendments

The 1990 amendments made a number of changes to the Clean Air Act, which had a direct impact on the issue of conformity between SIPs and transportation plans, programs, and projects. For the first time, the law specifically mandated that MPOs perform quantitative conformity assessments, using the type of air quality analysis that plaintiffs were urging the court to require the MTC to adopt. The amendments also touched on a number of the issues that had been dividing the parties in the lawsuit, including growth in emissions, use of emissions factors, and the need for evaluating individual highway projects and analyzing hot spots.

In enacting the 1990 legislation, Congress determined that there had been little guidance in the 1977 Clean Air Act on the issue of conformity and, as a result, local agencies had largely ignored it. No plan had ever been disapproved for lack of conformity, even in cities where the growth of mobile source emissions was a major factor in nonattainment.[40] As we have mentioned, transportation plans were traditionally developed to meet expected vehicle volumes without regard to the effect on air quality. In the 1990 amendments, Congress made it clear that transportation systems must meet the needs of both air quality and mobility.

The 1990 amendments emphasize reconciling the estimates of emissions from transportation plans and programs with the SIP, rather than simply providing for implementation of TCMs. A new SIP must eventually be prepared, reflecting the revised standards in the amendments. The law directed the EPA to issue new criteria for conducting conformity assessments by November 1991[41] and for states to revise their existing SIPs in accordance with these new regulations by November 1992.[42] Interim standards for conformity assessments were also set forth in subsection (3) of Section 176(c), which apply until the SIP revision is submitted and approved.[43] Although the amendments appeared to impose tighter standards, the MTC contended that the interim provisions superseded any existing conformity obligations under the Bay Area Plan and automatically validated all four highway projects. As far as the MTC was concerned, from then on it was to be governed solely by the 1990 amendments and no longer had to comply with the provisions in the 1977 act or the Bay Area Plan.

The Sierra Club insisted that the 1990 amendments did not replace the requirements in Appendix H because they contain a "savings clause" in Section 110(n), which ensures that existing SIP provisions remain in effect until the new SIP is approved.[44] In explaining the clause, the congressional conference report stated that projects must still comply with any existing conformity requirements and that the interim provisions were merely additional federal criteria that had to be met to qualify for federal approval or project funding prior to the revised SIP's being adopted.[45] Plaintiffs argued that the amendments preserved the original restriction on any MPO approving a TIP that does not conform to the applicable SIP, and that both Appendix H and Section 176(c) continued to govern those determinations. Thus, valid conformity assessments would still have to precede TIP approval for every project. Because the court had ruled the MTC's procedures violated federal law, no projects could now be approved based on their being included in those earlier transportation programs.

The MTC responded that the general "savings clause" did not bar application of the more specific transitional provisions, and that it would be unfair to single out the Bay Area for special treatment only because the court had disapproved its procedures, but other areas that had not been sued could continue to use their existing methods. MPOs in areas with far worse pollution problems had been using qualitative assessments for conformity determinations, and the FHWA had never objected. Until the SIP is revised to incorporate the new statutory definition of conformity, the interim provisions allow a project to qualify as conforming to the SIP under the grandfather clause in subsection (3)(B), which provides that for 12 months after the date of enactment of the 1990 amendments, projects are deemed to conform to the SIP if they came from an RTP or TIP that was "found to conform" within the past 3 years.[46] Congress, MTC argued, had recognized that most states' procedures were inadequate and intended to allow MPOs in nonattainment regions to rely for 1 year on their prior conformity assessments, however flawed, as they revised their SIPs.[47]

MTC argued that each of the four projects in question had been included in the previous 2 years' TIPs, which the FHWA had determined conformed to the Bay Area Plan, and that again the court had never ruled that these prior conformity findings were erroneous. Because there was no showing that any of the projects did not actually conform, they should be allowed to proceed, particularly because the 1990 amendments had, according to defendants, superseded the Bay Area Plan and rendered the issue moot. Should the final environmental reviews show there would be negative air quality impacts, the MTC promised it would recommend appropriate mitigation measures.

The Sierra Club, in turn, replied that the 1990 amendments were aimed at correcting noncompliance, not rewarding failure, and that Congress intended that every project conform, not that every project be exempted from conformity requirements. Even under the interim provisions, MTC could not justify its approval of the four projects. Subsection (3)(B) confirms that all highway projects must come from a conforming TIP, one meeting either the new criteria or the existing standards. Plaintiffs pointed out that because no conformity determination had yet been made for the TIPs with regard to the new standards, the MTC would have to qualify, if at all, under the grandfather clause. Because the court had already ruled MTC's procedures to be flawed, plaintiffs argued that no plans could have been "found to conform," so that there was no basis for applying the exemption. Besides, even if the FHWA had formally approved the earlier plans, the FHWA had merely accepted MTC's own flawed evaluation of conformity without conducting any independent analysis. Moreover, plaintiffs argued, the MTC had no idea how it could ensure conformity after the fact as part of the environmental review process, and charged that MTC was ignoring substantial risks to public health in preference to "short-term project economics."[48]

Carefully weighing the various arguments advanced by the parties, the court accepted the plaintiffs' statutory interpretation as most consistent with the purposes of the Clean Air Act. Declaring that passage of the 1990 amendments "underscores the importance of ensuring that transportation plans are consistent with SIP pollution control strategies," the court held that the Bay Area Plan remained in force, in view of the congressional intent to preserve existing SIP requirements and court decisions enforcing those requirements.[49] In addition, the savings clause specifically applied in the conformity context and during the interim period, so as not to "extinguish obligations independently imposed on MPOs by their respective SIPs."[50]

Inasmuch as the MTC had endorsed the Plan, it would hardly be "unfair" to the agency, the court concluded, to require it to meet its own commitments. The court also specifically rejected the MTC's contention that the grandfather clause covered the four projects in question. Because the TIPs they came from had been tested under faulty methods, the projects could not have been "found to conform" within the past 3 years.[51] The court also rejected the argument that the TIPs conformed just because the FHWA had already approved them.[52] In short, although the 1990 amendments created a limited exception during the interim period for projects that did not come directly from a TIP meeting the new conformity standards, it affected only projects coming from a TIP that was covered by a prior valid conformity assessment. Thus, the four projects would have to be delayed until they could undergo new conformity assessments, because they did not fall within the terms of the grandfather clause.

Congress, the court reasoned, had intended to strengthen conformity requirements, not reduce them, in light of the failures of earlier air pollution legislation, and to hold agencies to their existing obligations. The 1990 amendments made transportation plans and programs "a part of the pollution control strategy" for metropolitan areas.[53] To accept defendants' reading of the amendments, that any projects from a TIP adopted in the past 3 years were exempt from conformity requirements, would be to permit "backsliding" during the interim period. As the court explained, the grandfather clause was not intended to replace existing obligations but simply to maintain the status quo—in effect, it provided a floor for deciding conformity rather than a ceiling. Last, the court also rejected the MTC's argument that it was entitled to uniform treatment with other MPOs that had not had their procedures questioned, concluding that Congress had not intended to create a uniform national approach to conformity during the interim period but only to prevent increases in mobile source emissions that could undermine the objectives in the new legislation.[54]

Court Order

According to Judge Henderson, the conformity requirements were "of paramount importance in this nation's longstanding battle to achieve healthy air."[55] Finding it unacceptable that the MTC had been out of compliance with the Bay Area Plan since the May 7 Order, he issued an injunction against the MTC, prohibiting it from proceeding with any new highway projects without an adequate conformity assessment process. In crafting his order, the judge sought an approach that was "responsive to the violation yet pragmatically limited in scope."[56] He enjoined the MTC from approving projects until after new conformity procedures were approved by the court. He did not, however, require the MTC to vacate any of its prior approvals. Moreover, one of the four projects approved after the May 7 Order was deemed to be too far along to stop in light of the state's competing interests.[57] The remaining three projects, none of which had yet gone to bid, would have to be assessed as expeditiously as possible, using a procedure approved by the court, but in any event prior to commencing any construction. As for projects yet to be approved, in keeping with his flexible approach, the judge limited his remedial order to those projects that would increase highway capacity or interfere with the implementation of any TCMs. Such projects would have to demonstrate compliance before the MTC could grant final approval, except in cases of emergency or to increase the safety of existing highways, provided they did not "significantly" increase highway capacity.

The judge stressed that he was not halting or precluding any highway projects, merely "requiring MTC to carry out its federally mandated obligation to ensure that upcoming projects will improve and not pollute, our region's air before such projects become a permanent part of our landscape."[58] Nevertheless, the ruling had a major impact. For the first time, a court in the Bay Area had blocked a highway project because it did not conform with the state SIP. Attorneys for the Sierra Club felt that the court's decision enforcing this part of the Clean Air Act had national importance.[59] Residents and politicians throughout the Bay Area had to take note that even projects approved by the public to reduce congestion might become subordinated to air quality concerns. Highway planners, such as Greig Harvey, worried that the court's decision could affect how future highways were planned, including whether planners would have to consider the potential effects of a project on development before it could be approved.[60]

With the injunction question resolved, the court turned to the overall adequacy of the proposed conformity assessments, although it deferred a final ruling, in order to consider arguments on the possible impact of the 1990 Clean Air Act Amendments on the MTC's obligations. The court ordered the parties to file additional arguments on that issue. The MTC chose to revise its assessment procedures a second time in light of the 1990 amendments. Among the proposed modifications, it agreed to do modeling for the year 1997, in addition to 2010, reflecting new attainment requirements in the amendments and also changes in the California Clean Air Act.[61] Meanwhile, the MTC appealed the court's injunction order to the Ninth Circuit Court of Appeals and sought a stay of the injunction from the district court pending the appeal. These are discussed in the next section.

The parties hotly disputed the effect of the amendments on the MTC's legal responsibilities for conducting conformity assessments. The plaintiffs believed the amendments tightened the requirements on transportation agencies to consider air quality impacts, but the defendants viewed the new act as completely replacing the existing requirements. From the MTC's viewpoint, whatever new measures might be necessary to comply with the amendments should be a matter for decision by the EPA, which would eventually issue regulations interpreting the new law. Meanwhile, it felt the court should defer to the administrative process and not impose conflicting obligations on the local agency. Lawyers for the environmentalists again countered that the 1990 amendments did not relieve the defendants from any existing obligations under the Bay Area Plan because the savings clause in the new legislation preserved the MTC's commitments and the court's prior rulings.

Revised Conformity Assessment Procedures

Passage of the 1990 amendments caused the MTC to reevaluate its conformity assessment proposal in light of the new requirements. The major effect of the 1990 amendments was to extend the time for achieving the primary air quality attainment standards and to require some states to submit SIP revisions for meeting the new deadlines. As we noted in Chapter 1, the amendments categorize ozone nonattainment areas as Marginal, Moderate, Serious, Severe, or Extreme, based on their design values.[62] They also designate CO nonattainment areas as either Moderate or Serious.[63] The Bay Area is listed as a Moderate area for both pollutants. As discussed, a key part of the new law is that certain nonattainment areas must also adopt an emissions budget to demonstrate RFP milestones, attainment, or maintenance for particular years. These areas must submit revised SIPs reflecting the new standards.

The 1990 amendments also added a new general definition of conformity to Section 176(c), which endorsed the EPA's policy position that conformity findings for new highways should not result in violations of the clean air standards. Transportation projects, plans, and programs must not

1. cause or contribute to any new violation of any standard in any area,
2. increase the frequency or severity of any existing violation of any standard in any area, or
3. delay timely attainment of any standard or any required interim emission reductions or other milestones.[64]

This definition for conformity, MTC pointed out, was consistent with its existing policy and the definition that it had proposed and the plaintiffs had already accepted. In addition, however, the new legislation also explicitly requires states to perform an air quality analysis as part of the conformity process. In the future, the expected emissions from transportation plans and programs must be consistent with any emissions budget contained in the revised SIPs.[65] Thus, the 1990 amendments confirm the court's earlier finding that conformity is a "quantitative" issue. Finally, all TIPs must also still provide for "timely implementation" of TCMs consistent with the schedules in the SIP.[66]

Once the new SIP is approved, the new conformity assessment provisions take effect. Each RTP and TIP must be subjected to a regional emissions analysis, which ensures that no SIP violations will result from construction of the plan or program. This new provision in part addresses a long-standing dispute between the DOT and the EPA, alluded to earlier, over whether subsequent air quality analyses should be required for conformity determinations following the original SIP adoption. The EPA was also concerned, though, that highway projects were being changed or expanded after the RTP/TIP reviews without any further evaluations. Although the amendments retain the DOT's basic position, incorporated into its earlier regulations, that in general individual highway projects do not have to undergo a separate conformity assessment, they do deal with this problem. Under the 1990 amendments, a project now conforms only if

1. it comes from a conforming plan and program,
2. the design concept and scope of such project have not changed significantly since the conformity finding regarding the RTP and TIP from which the project derived, and
3. the design concept and scope of such project at the time of the conformity determination for the TIP were adequate to determine emissions.[67]

The amendments define interim conformity separately for plans and programs on the one hand and for projects on the other. During the interim period, no plans or programs may be approved unless they contribute to emissions reductions consistent with the new attainment standards for ozone and CO.[68] This ensures that states immediately begin meeting the new standards while the SIP revisions are being prepared. By November 15, 1996, all Moderate-and-above ozone nonattainment areas such as the Bay Area must achieve a minimum 15% reduction (net of growth) in volatile organic compounds (VOC) and oxides of nitrogen (NOx) from 1990 baseline conditions.[69] Moderate areas must also attain the primary standard for ozone by that date.[70] Moderate CO nonattainment areas are expected to meet the NAAQS for CO by December 31, 1995.[71]

During this interim period, projects are governed by the standards discussed earlier with respect to the injunction motion. To recap, projects must either come from a transportation plan or program meeting the preceding criteria or else qualify under the grandfather clause. In addition, projects in CO nonattainment areas must also eliminate or reduce the severity and number of air quality violations in the area substantially affected by the project.[72] Finally, to resolve the conflicting interpretations of the conformity provisions in the 1977 Clean Air Act, the 1990 amendments specifically authorized the EPA, with the concurrence of the DOT, to issue regulations establishing conformity criteria and procedures.[73] The MTC believed that the amendments superseded any conformity requirements in the 1977 act and that it was now up to the EPA to decide how conformity should be determined.

Although the agency made a few modifications to its proposal, the MTC generally felt that the new law supported the positions it had taken, including the lack of any need for individual project assessments or hot spot evaluations, and the propriety of using updated emissions factors in its computer model. On the other hand, clearly MTC would have to revise its attainment demonstration to comply with the new standards. Still, the agency took these new obligations as evidence that up until then neither Congress nor the EPA had ever expected MPOs to engage in the level of analysis demanded by plaintiffs with respect to conformity assessments. Given that the Bay Area Plan was written to comply with the old regulations, it could not have anticipated these "new" requirements. Therefore, it would be wrong for the district court to interpret the Bay Area Plan as having created any obligation to compare emissions estimates from the MTC's model of the region's transportation system with the emissions levels in the Plan used to select TCMs, or to require each individual project to undergo a separate conformity assessment. Plaintiffs, though, continued to insist that the Plan imposed more stringent requirements on the MTC that the 1990 amendments did not relax.

In response to the amendments, the MTC agreed to do a shorter term evaluation of its highway plans and programs. The agency proposed to model regional air quality for the year 1997, a date chosen to coincide with provisions in the California Clean Air Act. The MTC would generate projections for traffic and congestion levels in 1997, using MTCFCAST, and would rerun the Mode Choice and Traffic Assignment submodels to reasonable equilibrium. It would use STEP to evaluate Trip Generation and Trip Distribution sensitivity to travel times. It would not, however, reassess the POLIS regional population and employment projections for 1997, even if the "build" and "no-build" travel times were significantly different. These short-term predictions would be time-consuming and expensive, MTC argued, and besides, MTCFCAST and STEP already considered short-term changes in travel patterns and together captured the rearrangement of workers relative to job locations. Moreover, most major projects that might affect regional travel would not be completed before at least 1995, and ABAG's 1997 forecasts already included any possible land use changes from projects in the current TIP.

Based on this revised modeling, the MTC would compare the projections for each of the alternative transportation plans (highway capacity, transit capacity, and transit/highway blend) to the base year and select whichever was lower than both the current year conditions and the no-project alternative. It would make conformity determinations for the RTP and TIP for the years 1997 and 2010, although the agency adopted separate conformity tests for each analysis year. It would still not undertake any project-level conformity determinations, however, and would perform corridor assessments for only the year 2010.

RTP/TIP Conformity Assessments for 1997

MTC took the position that the revised Section 176(c) in the 1990 amendments required an air quality analysis focused on the new applicable attainment dates. Because the Bay Area expected to remain a Moderate nonattainment area for both ozone and CO, the key dates would be 1996 for ozone and 1995 for CO. For purposes of the 1997 assessment, the projects listed in the TIP and the RTP would be essentially the same. Using the MTCFCAST model, the MTC would calculate the expected emissions in that year from construction of these projects and compare the results to the new federal standards. Values for the interim years would be obtained through straight-line approximation. Under the MTC's proposal, the TIP and RTP would conform if the modeling showed that (a) by 1996, VOC emissions would decline from 1990 baseline levels by at least 15%,[74] and (b) by the end of 1995, projected emissions would be at or below levels necessary to attain NAAQS for CO [75] For these assessments, MTC would not look at individual projects or even specific corridors, because this would entail extensive modeling to create a 1997 "no-build" alternative for comparison, a step not required by the amendments (because the "build" scenario would be compared directly to the emissions standards).[76] The agency argued, though, that this new test did not cover long-range conformity.

Conformity Assessments for 2010

The MTC asserted that after passage, the 1990 amendments governed all subsequent conformity assessments. The interim requirements directed only that conformity be addressed with reference to emissions levels in the 1995-1996 period; it was silent on the question of conformity after that time. Although the Bay Area Plan had provided for maintaining air quality standards after the 1987 attainment date, the MTC believed that it did not explicitly mandate long-range conformity assessments. MTC's position was therefore that after the revised attainment dates, there were no applicable standards for long-term conformity, especially because the 1990 amendments seemed ambiguous on the need for a new maintenance plan.[77] Although the MTC felt that neither the Plan nor the amendments required long-term conformity, the agency agreed to continue modeling on a regional and subregional basis for the 2010 scenarios, on the assumption that Appendix H did require an assessment of the entire RTP and TIP, and most of the proposed projects would not be completed by 1997.

Although the 2010 assessment would be similar to that for 1997, it avoided the emissions budget approach applicable to the 1997 assessments. As long as the projected emissions in 2010 for the preferred "build" scenario were lower than the projected emissions from the "no-build" scenario and lower than the current base year emissions, then the TIP or RTP amendments would conform. This was substantially the same as the MTC's test for long-term conformity, discussed earlier, which had also rejected any ties to attainment level emissions. The MTC reiterated that the 1990 amendments did not link interim period conformity principles to the concept of long-term maintenance of NAAQS. Under these circumstances, as long as emissions from a plan or program were less than they would be if the projects were not constructed, then the RTP or TIP could neither "cause or contribute to a new violation," nor "increase the frequency or severity of any existing violation," nor "delay timely attainment of any standard."[78] Only if the 2010 "build" emissions were higher than the projected 2010 "no-build" emissions would the new attainment levels apply. In that case, the TIP or RTP amendments would still conform if the projected levels were lower than necessary to meet the standards by 1995 (for CO) and 1996 (for ozone).[79] If, however, the "build" levels were not only higher than the "no-build" condition but also higher than permitted by Section 176(c), then the MTC staff would review the TIP or RTP to decide what changes should be made to ensure that, on a long-range basis, they would meet the standards.

As for the effect of future transportation systems on land uses, the MTC reiterated its position that its model was capable of predicting changes in the geographical allocation of jobs and housing in the Bay Area and changes in trip patterns but it did not have to address the effects of new highway projects on overall levels of growth. The agency argued that in passing the 1990 amendments, Congress had specifically rejected proposed language that would have required MPOs to do an analysis of growth "likely to result" from transportation improvements.[80] The agency also insisted that the 1990 amendments clarified that it had no obligation to limit VMT or numbers of trips to the levels assumed in preparing the Bay Area Plan, because those sorts of requirements applied only to more heavily polluted areas.

In reassessing its modeling procedures in light of the changes to the Clean Air Act, the MTC also discovered that the 5% threshold for rerunning the POLIS projections for the 2010 model was a "hair trigger," which was not large enough to capture significant residential and job relocations because it would be exceeded by changes in travel times on the order of as little as 1 minute per day over a 20-year period. According to the MTC, this would be so insignificant as to not warrant the time and expense of redoing the demographic projections. Instead, the agency said it would rely on actual experience to recalibrate its model.

The Sierra Club conceded that the MTC's decision to model for the year 1997 mooted the debate over a single horizon year; however, it also raised new questions because, in their view, MTC had retreated from some of its earlier positions. As they had in their injunction motion, the plaintiffs again argued that Congress did not intend the 1990 amendments to lift the more stringent limits in current SIPs, but that existing SIP provisions were protected by the savings clause in Section 110(n). In short, they took the position that both the conformity commitments in Appendix H of the Bay Area Plan and the interim measures in the 1990 amendments applied pending approval of a SIP revision.

As we shall see in the next chapter, plaintiffs were generally concerned that updated emission estimates for the region in the 1990 inventory showed that, despite the new controls, pollution levels were much higher than originally projected in the Bay Area Plan. They also worried that the MTC was using the 1990 amendments as an excuse to abandon its efforts to attain the 1987 standards. From the plaintiffs' viewpoint, the revised assessment procedures were still flawed in allowing the MTC to find conformity even though new highway construction could lead to increased traffic and cause higher emissions. Their specific objections again focused on (a) the MTC's method for gauging conformity in the horizon years, (b) the MTC's refusal to account for increased traffic from changes in land use and development patterns resulting from new highway construction, and (c) the use of unproven tailpipe emissions factors.

Long-Range Conformity

Although the MTC's latest conformity assessment now addressed plan and program consistency in the short-term (1997), the Sierra Club felt it loosened the requirements for long-range conformity. Enactment of the 1990 amendments had established a link between conformity assessments and RFP emissions levels. Although both parties agreed that the MTC would have to comply with the interim requirements, they disagreed on how they should be interpreted. The plaintiffs insisted that the MTC's proposed new definition of conformity was inconsistent with the new Clean Air Act because it did not relate to the attainment levels in the Plan. By extending the original attainment dates, the Sierra Club reasoned, Congress had made those emissions levels again relevant to determining conformity.

Although the MTC's proposal limited emissions up to 1995 (for CO) and 1996 (for HC), in accordance with the interim provisions, it would also have permitted increased vehicle emissions after that time. As long as projected emissions for the "build" scenario stayed lower than both the "no-build" alternative and what the plaintiffs considered to be the "already excessive" present levels, long-term emissions could rise above the attainment levels in the Bay Area Plan and those required by Section 176(c). The Sierra Club maintained that inasmuch as the amendments set new attainment dates that the agency was not only obligated to meet but had agreed to achieve in its conformity proposal, to permit those emission levels to be exceeded after 1997 would indeed "cause or contribute to new violations" of the standards in violation of Section 176(c). Even if the 1990 amendments were somewhat ambiguous regarding interim rules for long-term conformity, plaintiffs argued, the savings clause preserved the Plan's provisions requiring a maintenance plan for air quality until the year 2010. In any event, until the EPA approved a SIP revision, the MTC could not exceed the emissions levels established in the Bay Area Plan.

The MTC, however, viewed the 1990 amendments as creating a whole new set of conformity requirements on MPOs, such as consistency with emissions targets, which superseded the old regulations. Thus, the agency believed that it was not under any obligation to maintain conformity beyond the revised deadlines. Even assuming that the Bay Area Plan remained enforceable under the savings clause, the MTC felt that plaintiffs were reading the new provisions back into the 1977 Clean Air Act and imagining that the 1982 Plan had imposed similar requirements on the agency that were then just being required of other MPOs. The MTC urged the court to defer to the EPA for administrative guidance on how long-term emissions comparisons should be made, because the matter was not clearly addressed in the 1990 amendments.

For its original 2010 analysis, the MTC had rejected tying emissions from the TIP or RTP to attainment-level emissions, a step the agency believed would have turned its conformity review into a new attainment demonstration. Now, the MTC complained, after initially accepting the MTC's conformity definition, plaintiffs had reversed their position to argue that projected emissions in 2010 should not be higher than attainment-level emissions. The agency believed that all the 1990 amendments did was extend the original attainment dates, not change the rules for long-range conformity. In the absence of EPA guidance, the agency argued, the court should not reject the previously agreed-on definition of conformity.

Traffic Growth

The 1990 amendments also caused the parties to renew their argument over modeling future traffic growth. The Sierra Club disputed the MTC's assertion that Congress had foreclosed consideration of excess growth in the 1990 amendments. Congress, plaintiffs insisted, had not intended to abandon the requirements in prior law that growth in emissions and growth due to vehicle traffic be addressed in each SIP.[81] In fact, it had strengthened the conformity requirements. They pointed out that the new interim conformity provisions for HC emissions in Section 176(c)(3) mandated that MPOs account for "any growth in emissions" after 1990.[82] The fact that Congress had eliminated other unnecessary language from the final version of the bill was seen as simply irrelevant.

The Sierra Club insisted that the 1990 amendments established that both the effects of changes in driving habits from improved capacity and the additional traffic from new development needed to be analyzed. Challenging the MTC's assertion that highway construction would not affect regional growth levels, the Sierra Club added that even if the POLIS model accurately reflected the future full "build" scenario, as the MTC claimed (by assuming construction of all long-term projects in the RTP), it would still not be appropriate to use the same population, jobs, and housing projections in connection with the "no-build" scenario, because without those projects fewer people might choose to live in the Bay Area. This is really the flip side of their earlier argument that the 2010 scenario understated traffic congestion. Here, plaintiffs asserted that the "no-build" scenario overstated traffic because it assumed the same higher population as if all the proposed highways were actually built. They pointed out that under existing local land use plans, certain housing developments might not be approved due to the lack of adequate transportation facilities.[83] In either case, the plaintiffs argued, the gap between pollution levels in the two scenarios was made to appear narrower than it should be so that the consequences of full build-out would seem less significant.

The MTC again responded that changes in highway facilities had little to do with total regional growth.[84] The new facilities in the TIP and RTP represented only incremental additions. There was no evidence that these projects would significantly increase the number of new residents or that not building them would deter any new population growth. Highway construction could, however, affect mode choices and route choices and, over the long term, possibly even automobile ownership and trip lengths. Taken together, these could have an impact on the location of housing and employment in the region; but as the MTC explained, the feedback loops in its MTCFCAST model accounted for these effects. Any new population moving into the Bay Area from outside the region that was prompted "solely by the availability of new transportation facilities" would, according to the MTC, amount only to a "tiny subcategory" of all new growth.[85]

The MTC explained that its future population projections from the POLIS model were reconciled with existing local land use plans and therefore represented "reasonably accurate forecasts of population and employment" in the region.[86] It insisted that its proposal would capture all new growth from persons moving into the region and the travel patterns of all commuters in the corridors generated by current and future residents, even if some projects in the RTP/TIP were not built. According to the MTC's experts, any developments that might not be permitted in a particular locality due to lack of freeway access would simply be shifted elsewhere in the region. The only aspect of travel that its modeling would not capture would be a hypothetical decrease in travel associated with a net loss of population over what might otherwise have been expected, resulting solely from a decision not to construct certain highway projects.

According to the MTC's consultant, the most important factors affecting population growth are (a) national and regional economic conditions and (b) demographic changes in the ethnic and cultural makeup of the population.[87] Land use policies, regardless of whether they are tied to transportation, have never been shown to restrict total regional growth. Unless regional growth controls are present, the agency argued, restricting development in one area would simply shift it to other areas so that the overall regional growth levels would not be significantly affected.

In reply, the Sierra Club pointed to a number of regional plans and reports as evidence of a connection between traffic congestion and transportation investment on the one hand and economic growth and regional development on the other. Even if regional population levels are not affected by particular projects, they argued, any "redistribution" of growth within the region was not in fact being captured by the ABAG forecasts, or at the very least was questionable in view of the MTC's uncertainty over the sensitivity of its model and the agency's decision to revise its POLIS "trigger" based on the model's future performance. This reallocation of traffic might affect congestion throughout the region, which could result in different levels of pollution. The problem would be especially acute on the subregional level in connection with local hot spot analyses, where decisions on whether to build particular projects could have a significant local impact even if they did not affect regional development.[88]

More critically, though, there was no assurance that any shifts in development from the combined effect of not constructing all projects in the RTP and TIP, as reflected in the "no-build" scenario, would necessarily be confined to the Bay Area. The extent and location of any replacement development, apart from that assumed in the full-growth "build" condition, would have to be taken into account for an accurate air quality assessment. As the plaintiffs' consultant explained, the uncertainty over whether and where replacement development would occur made the MTC's conclusions suspect:

> Where such a "build" alternative compares favorably to a "no-build" case only where replacement development is assumed to take place, however, then whether or not the "build" option would actually produce fewer emissions than the "no-build" case is uncertain. Given this uncertainty, MTC could not show that the "build" alternative would actually conform.[89]

In other words, plan and program conformity was dependent on the restrictive assumption that regional growth was entirely unaffected by transportation investment. Plaintiffs concluded that without empirical evidence to back this claim, the proposed assessment procedure could show only that conformity was *possible* but not that it was certain.[90]

ARB Tailpipe Standards

For the plaintiffs, the problem of potentially higher than anticipated vehicle use was still compounded by the MTC's use of the EMFAC 7e factors, which took credit for future tailpipe standards. Moreover, they pointed out that the EMFAC 7e factors had been developed by the California Air Resources Board, and the EPA had not yet approved them for use in California. MTC replied that using the current tailpipe projections conformed to the 1990 amendments' requirement that interim conformity determinations be based on the "most recent estimates of mobile source emissions,"[91] not outdated assumptions in the existing SIP. California has more demanding tailpipe standards than other states. According to the MTC, EMFAC 7e incorporates these standards and represents a vast improvement over previous emission factors in that it does not assume a fixed factor to estimate extra pollution caused by "cold starts," hot starts," and "hot soaks." Instead, it uses actual information on driving patterns in California to vary the pollution projections in more realistic ways for each category. But, plaintiffs insisted, the recent estimates requirement was designed only to ensure that MPOs would not understate vehicle emissions "by relying on outdated levels of vehicle use and congestion," but not so that the MTC could rely on emissions factors that were not part of the Bay Area Plan and had not been approved by the EPA.[92] The result of using these unproved factors would be unreasonably optimistic estimates of emission reductions.

In summary, the plaintiffs alleged that the defendants' proposed plan and program conformity assessment would produce artificially low estimates for future increases in air pollution by (a) underestimating emissions associated with the "build" scenario, overestimating emissions from the "no-build" scenario, or both; (b) not accounting for traffic growth due to highway expansion; and (c) applying untested emissions factors to those increased traffic levels.

Court Ruling on
Revised Conformity Assessment

At this point in the lawsuit, in addition to disputing the extent to which freeway construction might induce new growth, the parties were also still deeply divided over whether individual projects were subject to air quality assessments and over the appropriate way to address CO hot spots. In March 1991, after reviewing the extensive record in the case, considering the arguments of the parties, and consulting with Dr. Wachs concerning technical transportation issues raised by the case, the court completed its analysis of the revised conformity assessment procedures and issued its decision. The court held that, with one exception, the MTC's latest proposal adequately satisfied its conformity obligations under the Bay Area Plan and the interim conformity provisions of the 1990 Clean Air Act Amendments.[93] The court saw the 1990 amendments as supplementing the agency's obligations under the Plan by providing substantially more detailed information on ensuring plan, program, and project conformity but not replacing them.

In making its ruling, the court maintained its flexible approach, recognizing the limited role played by the courts in overseeing government agencies. Although the court agreed with plaintiffs that it had an obligation to independently decide whether the agency had complied with the SIP (because the issue of compliance required "complex judgments in highly technical areas") the court chose to follow a "reasonableness" standard, showing deference to agency expertise and discretion.[94] The court called MTC's proposal a "quantum leap forward" from the MTC's previous practices and described it as a "sophisticated, quantitative approach utilizing advanced computer modeling techniques."[95] With respect to most of the technical issues raised by the plaintiffs, the court accepted the MTC's positions. It held, for instance, that the MTCFCAST model did an adequate job of assessing the impact of freeway improvements on the distribution of growth in the region and that, with some restrictions, the MTC could use the most current emissions factors in its calculations. As for the legal questions raised, the court again tended to agree with the MTC. The court ruled that the MTC's long-range conformity assessments were adequate and that it did not have to account for any potential growth due to highway expansion. It also determined that neither the Bay Area Plan nor the 1990 amendments required separate conformity assessments for individual projects, although the MTC was obligated to perform some preliminary analysis to identify hot spots to satisfy RFP for CO at the subregional level.

Long-Range Plan and Program
Conformity Requirements

As a preliminary matter, the court held that the MTC's definition of conformity for the year 2010 was not unreasonable, particularly given the lack of any regulatory guidance from the EPA in this area. In effect, it permitted the agency to find plans and programs in conformity if the long-range modeling showed both no increase in emissions over present levels and that emissions for the "build" scenario would be lower than those for the "no-build" case, or at least less than the level associated with achieving RFP by the interim year 1997. The court did not believe this to be "clearly contrary" to the Bay Area Plan or the 1990 amendments. Subsequently, the EPA has issued new regulations concerning the need to establish long-term conformity as part of the interim period assessments. These developments are discussed in the final chapter. The court then turned to the issue of regional growth impacts.

Regional Growth Impacts

To summarize the parties' positions at this stage, both sides agreed that ABAG uses a regional growth model to project the aggregate future population and housing stock on the basis of factors such as birth, death, and migration rates and economic trends, but that the modeled rate of growth is not explicitly a function of the region's transportation system. Regional population and housing unit totals are determined prior to and independently of any transportation plans. Highway systems are then designed to match the expected traffic, based on those projections, and so therefore would not in themselves stimulate significant additional growth.

Plaintiffs held that if the "no-build" scenario should occur, at least some of the projected regional housing would not be built and the total population would be less than anticipated. In short, some growth that would be attracted to the Bay Area in the presence of a good highway system would not occur at all. Consequently, the conformity assessment should address two aspects of growth. First, the "no-build" alternative should reflect a decrease in total projected regional growth, and second, it should be associated with a differ- ent spatial distribution of the population than the "build" alternative. In the comparison between "build" and "no-build," the advantages from the "build" proposal would therefore have to be greater to justify proceeding with the project.

The MTC insisted, to the contrary, that there is no scholarship that empirically associates total regional growth rates with the quality of the transportation system or with levels of traffic congestion. The POLIS model does, however, *distribute* the population projections among zones in the Bay Area as a function of the extent and quality of the transportation system. That is, areas having good accessibility are considered more attractive to growth than more remote areas. Thus, new highways are modeled to attract a larger proportion of the total regional growth to their vicinity than would otherwise locate there if those highways were not built.

MTC acknowledged to the court that it could certainly model a different spatial distribution of total growth under the "no-build" case than the "build" case but objected to reducing growth under the "no-build" case, because there were no models available to do that and no empirical basis for estimat- ing how much growth would be decreased. Indeed, the agency asserted that there was ample unused capacity under existing zoning to accommodate growth in other areas of the Basin, even if some planned highway projects were not completed.

The court concluded that MTC's decision not to analyze capacity-related growth was not unreasonable. Although the court acknowledged that plain- tiffs' concerns "may well have validity" and noted that the EPA might ultimately require them to be addressed in some manner, nevertheless it found that no clear link existed between highway capacity and regional growth in developed regions such as the Bay Area. It also agreed that no models were presently available that could estimate such effects even if they were present. Moreover, it found there was no mandate either in the Bay Area Plan or in the 1990 amendments to consider such changes. There was, however, evidence that highway investment could affect the distribution of growth, but the court concluded that the MTC's new models adequately took this into account.

Emissions Factors

The court was also not persuaded that the emissions factors used in preparing the Bay Area Plan should be treated as enforceable TCMs or that either the Plan or the Clean Air Act required MTC to use outdated information and assumptions. Indeed, the court agreed that the 1990 amendments specifically require the MTC to use "the most recent estimates of mobile source emissions" during the interim period before the new SIP is approved. The court felt that accurate modeling of air quality impacts demanded the use of the most current estimates of vehicle trips, VMT, and level of emissions produced by average vehicles on the road, including detailed estimates by type of vehicles, weighted by their proportion in the fleet. MTC was therefore reasonable in using the emissions factors that were most applicable to California, not the most stringent or those that demanded the most control measures. Still, it was also reasonable for the plaintiffs to insist that MTC use only factors approved by the EPA. Therefore, the court ruled that the MTC could use the ARB's latest emissions factors; however, once the EPA concluded its review of the EMFAC 7e factors, MTC would have to abide by the EPA's determination.[96]

Model Validation

The court recognized that each of the models (MTCFCAST, POLIS, and STEP) had been individually validated but not the entire modeling approach. Because validation of the combined models would be costly and could take several months, the court ruled it was reasonable for MTC to begin using the new procedure immediately, making further validity evaluations based on actual experience.[97] As to the remaining disputes, with some minor exceptions, the court again by and large approved the MTC's efforts.

Project-Level Analysis

The Sierra Club continued to insist that Appendix H required the annual TIP assessments to include an individual assessment of "major highway projects" and that this provision was protected by the saving clause in Section 110(n) of the 1990 amendments. This analysis was necessary, in plaintiffs' view, to ensure that individual projects would not increase pollution and thus exacerbate existing violations or delay attainment. Project-level analysis was also needed to determine whether to delay particular projects under the Contingency Plan because they would significantly increase pollution. As noted earlier, for the 1990 to 1997 period, the MTC did not propose to do modeling at even the corridor level, much less for individual projects.

The MTC argued that the 1990 amendments settled whether project-level conformity assessments were required. The agency insisted that the amendments supported its position, consistent with existing FHWA regulations, that project conformity was solely a function of plan and program conformity. According to the new law, once the SIP revision is approved, projects conform if they are from a conforming RTP or TIP, the project is sufficiently well-defined in the plan or program to determine emission levels, and the design concept and scope of the project do not undergo any substantial changes after the conformity assessment is made.[98] Only projects that fail to meet these tests must undergo a separate conformity analysis to prove that projected emissions from the project do not cause the plans and programs to exceed the emission reduction schedules in the SIP.[99] The MTC believed that this confirmed that the 1982 Plan could not have required project-level conformity assessments.

The MTC insisted it should be allowed to run its 1997 model without considering individual projects, even on a corridor basis, because the validity of project assessments in the short term could be affected by the fact that some projects might be only partly completed by 1997. In any event, experts for the MTC argued, a conformity analysis of projects should not be undertaken out of context, particularly given the difficulty of measuring regional impacts with any degree of accuracy.[100] Although plaintiffs contended that the MTC's proposal would allow some projects to "slip through the cracks," the MTC's experts felt this was unlikely as a practical matter, because all projects would still have to go through a thorough environmental review before receiving final approval.[101]

Because it is the *region* that must meet the standards, clearly some individual projects will make air quality worse and some will make it better. Should the area as a whole fail to maintain RFP, and the EPA require a SIP revision, the Contingency Plan would be triggered, requiring a review of projects for possible delay. This would not only ensure that projects would not contribute to a violation of air quality standards, the MTC concluded, but this additional step would also be unnecessary if each individual project had already undergone a separate analysis to show it did not adversely affect air quality.[102] Thus, project-level reviews were neither required nor necessary.

The court agreed with the MTC that the "plain language" of the Bay Area Plan supported determining the conformity of the TIP as a whole and that neither the Plan nor the Clean Air Act required any project-specific analyses, noting that the agency had also raised reasonable technical grounds for not undertaking them. In addition, the court considered modeling all projects in corridors for 2010 to be superior to individual project analyses, because it would allow for evaluating the synergistic and cumulative effects of projects on one another. Under the 1990 amendments, transportation projects would conform as long as they came from a conforming transportation plan and program, provided that the design concept and scope of the project did not change significantly. If in the future, the court reasoned, the region did not meet RFP, then the contingency provisions would apply and any projects with significant air quality impacts would have to be delayed, thus minimizing the need for initial project-level reviews.[103]

CO Hot Spots

The last point of contention concerned MTC's proposal to defer hot spot analyses until the environmental review stage of the projects in the affected areas, rather than as part of the annual RTP and TIP conformity process. As already noted, carbon monoxide (CO) pollution differs from other regulated pollutants because it is most harmful at specific locations such as busy urban intersections surrounded by high-rise buildings, where there is heavy traffic and little opportunity for the emissions to be mixed with fresh air.

Plaintiffs objected to MTC's decision not to review projects for CO hot spots, saying it was contrary to Appendix H, which stated the air quality impacts of RTP amendments and major TIP projects must be determined at the time of the RTP amendment or TIP adoption. They also asserted that the interim plan and program consistency provisions in subsection (3)(A) of Section 176(c) required the MTC to demonstrate sufficient emissions reductions to achieve the CO standards at the hot spot level. As noted, MTC proposed to make this evaluation for projects in CO hot spot areas later, at the environmental review stage.

The MTC responded that it did not need to analyze local impacts as part of the plan or program conformity assessment because subsection (3)(B) of Section 176(c) now governs all project-level conformity reviews during the interim period. Although projects in CO nonattainment areas must "eliminate or reduce" violations of CO standards in the vicinity of the project, the section also provides that the CO issue may be evaluated any time prior to final approval.[104] The MTC reiterated that when projects are included in the TIP and the RTP, they are not at a state of design detail that would permit identification of CO hot spots, but that these would be picked up in the environmental review process.

As the MTC pointed out, at the time that a project is included in the TIP, the project is not yet fully designed or engineered. Typically, it is identified simply as an "x-lane highway from point a to point b," and its particular alignment is not determined, nor is the configuration of on- and off-ramps, overpasses, and so on. The extent to which a project may create a CO hot spot depends on its detailed design, and therefore the environmental review is the most appropriate time at which to assess the extent to which it may create hot spots. The 1990 amendments recognize this by requiring projects to specifically demonstrate that they will eliminate or reduce the severity of CO violations, but allowing that showing to be postponed to the environmental review stage, where the project will be more completely described.

On the other hand, environmental documents are often poorly drafted, and agencies typically become more committed to the construction of their projects as the details are finalized. There is a tendency of highway agencies to insist, at the preliminary design stage, that environmental review is better done later and then to insist later that so much time, energy, and financial resources have already been spent on the project that to slow it down or revise it for environmental reasons would be a waste of public resources. The court realized that it is important not to let projects get so far along without any hot spot assessments that they become politically impossible to modify or even stop. It therefore ruled that the MTC's decision to delay hot spot analyses was not reasonable.

Although the 1990 amendments allowed for delaying CO assessments with respect to individual projects, the court noted that there is no such exception for *program* conformity evaluations. The court acknowledged that when projects are first included in a TIP, they may lack specific design details to perform such an analysis. Nevertheless, the court held that to satisfy its plan and program conformity responsibilities for CO, the MTC would at least have to do some preliminary qualitative analyses to identify projects in the TIPs with a significant potential for creating or contributing to hot spots. This, according to the court, would give a "fuller view" of the TIP's potential impact on emissions. More rigorous hot spot analyses could be performed later, at the environmental review stage, when the projects were better defined.[105] To not further delay the conformity process, the court ordered MTC to submit a proposal for undertaking these preliminary hot spot analyses, which would be incorporated into future conformity determinations.[106]

Finally, the court also heard the MTC's motion for a stay of the injunction at this time. The MTC contended there was no evidence that any of the enjoined projects would cause any harm and that each would likely pass under valid procedures, but that delaying construction would cost taxpayers upward of $300,000 a month. In addition, six major capacity-increasing projects were in process of being submitted, which might be jeopardized, particularly if the plaintiffs appealed the court's ruling on the validity of the MTC's new conformity procedures. These projects would also be subject to increased inflation costs due to any further delays. Bay Area commuters, the MTC asserted, had a strong interest in eliminating congestion as soon as possible, which outweighed any remote environmental harms.

Judge Henderson turned down the request for a stay. Although noting that the legal issue of whether the grandfather provision barred the injunction was an important question of first impression, he concluded that the public interest in reducing traffic a little sooner did not justify granting the stay, because the likelihood of an appellate reversal was not strong, given the express savings clause, and because the MTC would not suffer any irreparable injury from delaying the projects a short time until they could be tested under the approved procedures. In fact, the public had a strong interest in enforcing the MTC's conformity obligations under the Clean Air Act and the Bay Area Plan, rather than accepting the MTC's own assurances that the projects would probably conform:

> Notably, the 1990 Amendments to the Clean Air Act stress that the conformity requirement is a critical component of Congress' strategy to reduce air pollution and achieve healthy air. . . . Without a valid conformity assessment, there can be no such assurance. MTC is, in effect, asking the Court to assume precisely what Congress intended the conformity procedures to determine.[107]

In an unpublished opinion, the appellate court subsequently affirmed Judge Henderson's original injunction order, except those parts that became moot once the district court approved the MTC's revised conformity assessment procedures. The Court of Appeals concluded that, faced with the MTC's ongoing violations of federal law, the district court had crafted an appropriate equitable remedy.[108]

Conclusion

Despite the court's decision approving the MTC's revised procedures, the plaintiffs had accomplished a great deal, from their standpoint. By filing the suit, they had forced the MTC to reassess its highway construction program and adopt new modeling procedures to evaluate the impact of proposed roadways. The agency would have to employ those procedures in evaluating not only existing projects but also all its future highway plans and programs. The district court had also endorsed the plaintiffs' main contention that the transportation system has a potential impact on the distribution of growth in a region, which can affect congestion and air pollution levels. The MTC would have to consider how its plans and programs would affect future traffic patterns and the potential impact of new highways on development in the region served by those facilities. Although the court had refused to require the MTC to consider whether new roadways would induce additional growth in the region, the plaintiffs had at least made an issue of the relationship between transportation and land use.

The plaintiffs could take heart that, whatever their shortcomings, the procedures developed by the MTC in response to the lawsuit could serve as a standard for other areas in the future. Although the court did not base its ruling directly on the 1990 amendments, it clearly took support from the fact that the new legislation mandates conformity procedures quite similar to those the court felt were dictated by the Bay Area Plan.

Having disposed of the conformity question, the final issue left for the court to decide was whether the draft contingency measures the MTC had submitted satisfied the requirements of the Bay Area Plan. Specifically, the court addressed whether the new proposed batch of TCMs would produce sufficient reductions in air pollution to make up for the deficiencies identified by the court in its rulings on the summary judgment motion, which were described in Chapter 2. In doing so, the court would have to consider the fact that the initial estimates derived from the MTCFCAST modeling showed the region to be even farther "off the RFP line" than expected. The next chapter discusses both the parties' competing contentions regarding the MTC's use of its new model to project potential air quality improvements from the proposed TCMs and the resolution arrived at by the court.

Notes

1. DOT/EPA, *Clean Air Through Transportation: Challenges in Meeting National Air Quality Standards* (August 1993): 30.

2. POLIS stands for "Projective Optimization Land Use Information System." The POLIS model generates growth projections for 114 separate analysis zones, which are then manually allocated to the approximately 1,200 census tracts in the nine-county region. It then distributes the resulting projections to cities within the region. Although the model does not rely directly on changes in the transportation system to predict land use patterns, it is sensitive to changes in travel times. At the time of the lawsuit, the POLIS model did not consider the availability of transit alternatives when measuring travel times, although a "Transit Capacity Alternative with Land Use Sensitivity" was being developed by the staff. Declaration of Charles L. Purvis in Support of MTC's Revised Conformity Assessment Procedures, July 3, 1990, ¶¶ 19-22.

3. STEP divides the Bay Area region into 34 separate districts and computes the number of trips that will be made from any one district to any other. It also projects the number of work and nonwork trips, the number of transit versus vehicular trips, and the number of drive alone, shared ride with two occupants, and shared ride with three or more occupants vehicular trips for each household. The results for the 34 districts can be allocated to each of the 700 zones in the MTCFCAST model. Declaration of Greig W. Harvey in Support of MTC's Revised Conformity Assessment Procedures, June 30, 1990 [hereinafter Harvey Decl.], at ¶¶ 32-33.

4. *Citizens for a Better Environment v. Deukmejian,* 34 Environment Reporter Cases (ERC) 1689, 1695 (N.D. Cal. 1991).

5. "Major projects" were defined as those that "increase the capacity of the highway system through (1) addition of new highways where none existed before, (2) significant widening or addition of one or more lanes to an existing highway, or (3) improvement of traffic flows through addition of ingress and egress facilities on or between existing highways." The MTC's Memorandum of Points and Authorities in Support of Revised Conformity Assessment Procedures, July 3, 1990, at 9-10.

6. At first, the MTC did propose to compare the results from the 2010 projections with the estimated levels in the Bay Area Plan *associated with attainment of the air quality standards by 1987* and to require additional TCMs to offset any difference before a conformity finding could be made. In other words, the TIP and RTP would conform if the projected levels in the "build" scenario in the year 2010 were less than or equal to the RFP targets for 1987 once the mitigation measures were factored in. Although plaintiffs agreed with the practice of comparing emissions levels, they objected to the fact that the procedure tested the network in the horizon year against 1987 standards and so allowed conformity to be postponed for several decades. They also complained that the agency was relying on untried and unenforceable mitigation measures, which were not part of the Bay Area Plan, to establish conformity rather than delaying or canceling any highway projects in the event that the projections exceeded allowable pollution levels. In its

revised proposal, the MTC dropped the direct emissions level comparisons and reliance on additional TCMs, arguing that it would have, in effect, converted the conformity determination into a new demonstration of attainment, which would have required a formal SIP revision.

7. Metropolitan Transportation Commission's Reply Memorandum in Support of Revised Conformity Assessment Procedures, September 12, 1990, at 6-7 [hereinafter MTC's Reply Memo].

8. See former 23 C.F.R. § 770. 9(a) (originally published at 46 Fed. Reg. 8429 (January 26, 1981)).

9. Because all parties agreed that the MTC should not include additional TCMs in its assessment to offset any amount by which emissions in the "build" scenario exceeded attainment level mobile source emissions (see note 6 supra), requiring the MTC to establish that emissions levels from the transportation system were at or below attainment levels could prevent the agency from approving any plan, including the "no project" option. Therefore, MTC argued, the conformity analysis should avoid any ties to attainment levels.

10. Sierra Club's Supplemental Memorandum in Opposition to MTC's Amended Conformity Proposal, October 11, 1990, at 24 [hereinafter Sierra Club's Supp. Memo].

11. Plaintiffs reconsidered their position in part based on the new conformity definition being proposed by the EPA in the amendments to the Clean Air Act being considered by Congress. See text accompanying note 64 infra. In addition, the MTC had originally proposed to allow plans and programs to increase emissions beyond the RFP targets as long as additional mitigation measures were adopted to offset any difference. See note 6 supra. Inasmuch as plaintiffs had objected to this proposal, they seemed to agree that a test contemplating no increase in emissions was preferable

12. The MTC claimed its method was more advanced than others in use because the MTCFCAST, STEP, and POLIS models captured more of the "direct, secondary and tertiary effects of highway capacity expansion." Supplemental Declaration of Greig Harvey in Support of MTC's Revised Conformity Assessment Procedures, September 12, 1990, at ¶ 3 [hereinafter Harvey Supp. Decl.]. MTC's consultant also noted that it would be a useful prototype for conformity assessments under the 1990 amendments to the Clean Air Act.

13. According to the MTC, long-term projections actually give a penalty to highway construction because of the negative air quality impacts caused by location and land use shifts from highway expansion. Modeling for interim years would not show as great an impact on increased capacity and congestion. Also, short-term modeling benefits from lower population projections, reduced VMT and number of trips, and increased average speeds due to reduced congestion, but disregards long-term travel behavior and land use shifts that might offset air quality benefits. MTC's Reply Memo, supra note 7, at 29.

14. Sierra Club's Supp. Memo, supra note 10, at 16-18. Plaintiffs suggested that some effects from additional highway projects would clearly be felt in the "medium term." Id. at 16, n.17.

15. CBE's Memorandum of Points and Authorities in Support of Motion for Summary Judgment Against the MTC, September 24, 1990, at 19.

16. The defendants were quick to point out that this argument was similar to that used by the MTC itself with respect to the Contingency Plan—that the court could not order a modification to the Bay Area Plan because it was up to the EPA to request a formal SIP revision due to "incorrect assumptions of growth rates of travel demand." MTC's Reply Memo, supra note 7, at 4. See former 23 C.F.R. § 770. 9(e) (originally published at 46 Fed. Reg. 8429 (January 26, 1981)).

17. MTC's Reply Memo, supra note 7, at 8.

18. Plaintiffs also complained that the MTC did not run its "no-build" scenario iteratively to equilibrium and thus did not consider the effects of the lack of new highway capacity on either travel demand or regional development. Therefore, although it provided a useful benchmark figure, it did not realistically predict actual pollution because it overstated future congestion levels. The MTC explained that in its original proposal, the projections for the 2010 "no-build" scenario were simply used to test whether changes in travel times were sufficient to trigger a reevaluation of regional growth projections, therefore the model was not run to equilibrium. See note 6 supra. Under the revised proposal, the MTC would be comparing "build" and "no-build" levels directly, rather than comparing "build" conditions to attainment levels, so the MTC agreed to run the "no-build" model to equilibrium. If there was a 5% difference in travel times, MTC would request new projections from ABAG. If the POLIS model showed a 5% or greater shift in population location for any corridor, then the MTC would rerun MTCFCAST and STEP in an iterative manner for the "no-build" alternative.

19. Declaration of Peter R. Stopher in Support of Sierra Club's Objections to the MTC's Proposed Conformity Assessment Procedures, August 20, 1990, Ex. B, "Traffic Congestion & Capacity Increases," p. 2 [hereinafter Stopher Decl.].

20. Id. at Ex. B, p. 26.

21. Id. at ¶ 8.

22. Id. at ¶ 9 & Ex. B, pp. 26-31.

23. See *League to Save Lake Tahoe, Inc. v. Trounday,* 598 F.2d 1164, 1174 (9th Cir. 1979), *cert. denied,* 444 U. S. 943 (1979). In that case, the plaintiffs sought review, under the Clean Air Act, of a state agency decision to issue a development permit. Because, under the applicable SIP, the agency director could not issue the permit if he determined that the activity would prevent the attainment or maintenance of NAAQS, the plaintiffs argued that issuing the permit was an abuse of discretion, because the technical analysis on which the action was taken "did not take into account the situation that would occur under the most adverse meteorological conditions and failed to consider CO levels within the project areas." Id. at 1168. The Ninth Circuit Court of Appeals viewed the case as an attempt to obtain judicial review of a discretionary administrative decision and denied relief because the specific procedural requirements in the SIP for obtaining the permit had been complied with. Relying on dicta in the opinion of the court stating that "discretion should properly repose in the responsible state official to establish such computer models and analysis as they deem appropriate," the MTC contended that the Clean Air Act did not require it to quantify vehicle emissions or use any particular type of model, and therefore under *Trounday* it had discretion to choose its own approach. The Sierra Club responded that *Trounday* was inapplicable and that the MTC's interpretation of its SIP was not entitled to judicial deference based on the holding in *American Lung Association v. Kean,* 670 F. Supp. 1285, 1291 (D. N.J. 1987), *aff'd,* 871 F.2d 319 (3rd Cir. 1989), that federal courts have discretion to fashion appropriate remedies for SIP violations. MTC responded that in that case, the State of New Jersey had denied its SIP commitment altogether and the court had ordered it enforced according to its terms, whereas here MTC did not deny that Appendix H required it to do a conformity assessment, just not in the way plaintiffs wanted.

24. MTC's consultant contended that there was no "practical, tested model" that could capture the effects of transportation infrastructure on "macroeconomic determinants of regional economic growth and demographic change." Harvey Supp. Decl., supra note 12, at 7.

25. Harvey Decl., supra note 3, at ¶ 27a.

26. EPA/DOT, *Procedures for Conformance of Transportation Plans, Programs and Projects with Clean Air Act State Implementation Plans* (June 1980).

27. Air Quality Conformity and Priority Procedures for Use in Federal-Aid Highway and Federally Funded Transit Programs, 46 Fed. Reg. 8429, 8430 (January 26, 1981) (codifying former 23 C.F.R. § 770. 9(c), *removed by* 57 Fed. Reg. 60,728 (December 22, 1992)).

28. A further review was needed only if (a) a supplemental environmental impact statement (EIS) significantly related to air quality was undertaken, (b) the EPA requested a SIP revision, or (c) major steps toward implementation of the project had not commenced within 3 years of the date of approval of the EIS. Id. (former 23 C.F.R. § 770. 9(d)). If a SIP revision was requested, the FHWA would temporarily refrain for 12 months from authorizing any project listed for delay under the contingency provisions triggered by the SIP revision unless it was exempt from sanctions under Section 176(a) of the Clean Air Act. Id. (former 23 C.F.R. § 770. 9(e)(2)). After the 1987 deadline had passed, the DOT proposed to review projects only if there were sufficient changes to make a supplemental EIS necessary. 53 Fed. Reg. 35,178, 35,181 (September 9, 1988).

29. Sierra Club's Supp. Memo, supra note 10, at 5. The plaintiffs believed that the test of conformity for highway projects should be that "the activity would not result in emissions greater than existing levels in future years without the project or program." Id. at 6.

30. Approval of 1982 Ozone and Carbon Monoxide Plan Revisions for Areas Needing an Attainment Date Extension, 46 Fed. Reg. 7188, ¶ G (January 22, 1981). In this 1981 SIP approval policy, the EPA stated that "Section 176(c) requires that any increase in emissions from mobile or stationary sources that result directly or indirectly from the construction and operation of a federal facility must conform to the SIP," adding that the SIP revisions should identify, to the extent possible, the direct and indirect effects of all major federal actions planned during the period covered by the SIP to enable state and local governments to more quickly and easily evaluate subsequent federal conformity determinations.

31. Conformity of Federal Actions to State Implementation Plans, 45 Fed. Reg. 21,590, 21,590-1 (April 1, 1980). The EPA proposed to promulgate regulations requiring each state to establish criteria and procedures to ensure conformity of federal actions with the SIP, including a requirement of consistency with the state's approach for demonstrating RFP during the period prior to attainment of the NAAQS. The EPA's position was that the increased emissions had to be accommodated in the SIP emissions growth increment for the nonattainment area; however, no further rule-making action was taken at that time.

32. Sierra Club's Supp. Memo, supra note 10, at 24.

33. MTC's Supplemental Reply Memorandum of Points and Authorities in Support of Revised Conformity Assessment Procedures, October 22, 1990, at 10.

34. The projects were (a) I-680 and Highway 24 interchange reconstruction in Walnut Creek, (b) addition of two HOV lanes to a portion of I-680 through the San Ramon Valley in Contra Costa and Alameda counties, (c) widening a part of I-880 in Santa Clara and Alameda counties by adding two mixed-flow lanes, and (d) upgrade of State Route 237 in Santa Clara County to freeway status and addition of two new part-time HOV lanes west of the I-880 interchange.

35. 42 U.S.C.A. § 7604(a) (West 1983 & Supp. 1994) ("The district courts shall have jurisdiction, without regard to the amount in controversy or the citizenship of the parties, to enforce such an emission standard or limitation"). See also, *California Tahoe Regional Planning Agency v. Harrah's Corp.*, 509 F. Supp. 753, 761 (D. Nev. 1981); *California Tahoe Regional Planning Agency v. Sahara Tahoe Corp.*, 504 F. Supp. 753, 769 (D. Nev. 1981).

36. As such, the MTC chose to treat plaintiffs' motion as one for a *preliminary* injunction and argued that the traditional equitable considerations would apply. The MTC insisted that plaintiffs had not shown a statutory violation or even a probability of success in proving that the TIP did not conform, nor any evidence of irreparable harm. Plaintiffs took the position that the court's ruling on summary judgment had settled the liability issue and that the Clean Air Act mandated injunction as the appropriate enforcement remedy and that, in any event, the balance of harms tipped in favor of plaintiffs.

37. Section 176(c) of the 1977 Clean Air Amendments provided in part, "The assurance of conformity to such a plan shall be an affirmative responsibility of the head of such department, agency, or instrumentality." The language is carried forward in the 1990 amendments. See 42 U.S.C.A. § 7506(c) (West 1995). Defendants argued this vested discretion in the FHWA to make conformity determinations. Indeed, MTC also argued that there was no affirmative duty for MPOs to even carry out conformity assessments inasmuch as Section 176 imposed the duty to ensure conformity on the FHWA.

38. *CBE v. Deukmejian,* 34 ERC 1592, 1595 (N.D. Cal. 1990).

39. Id. at 1596-1597.

40. 101st Cong., 2nd Sess., 136 Cong. Rec. S16956 (daily ed. October 27, 1990).

41. 42 U.S.C.A. § 7506(c)(4)(A) (West 1995). The regulations were to cover at a minimum (a) the procedures for consultation between MPOs, the DOT, and state and local air quality agencies and state departments of transportation; (b) the frequency of conformity determinations; and (c) how conformity determinations would be made with respect to maintenance plans. Id. § 7506(c)(4)(B).

42. 42 U.S.C.A. § 7506(c)(4)(C) (West 1995). Section 176(c)(4)(C) required each state to submit a SIP revision by November 15, 1992, that included criteria and procedures for assessing conformity of any covered plan, program, or project.

43. Section 176(c)(3) provides

Until such time as the implementation plan revision referred to in paragraph (4)(C) is approved, conformity of such plans, programs, and projects will be demonstrated if—

(A) the transportation plans and programs—

(i) are consistent with the most recent estimates of mobile emissions;

(ii) provide for the expeditious implementation of transportation control measures in the applicable implementation plan; and

(iii) with respect to ozone and carbon monoxide nonattainment areas, contribute to annual emissions reductions consistent with sections 7511a(b)(1) and 7512a(a)(7) of this title; and

(B) the transportation projects—

(i) come from a conforming transportation plan and program as defined in subparagraph (A) or for 12 months after November 15, 1990, from a transportation program found to conform within 3 years prior to such date of enactment; and

(ii) in carbon monoxide nonattainment areas, eliminate or reduce the severity and number of violations of the carbon monoxide standards in the area substantially affected by the project.

With regard to subparagraph (B)(ii), such determination may be made as part of either the conformity determination for the transportation program or for the individual project taken as a whole during the environmental review phase of project development.

42 U.S.C.A. 7506(c)(3) (West 1995).

44. Section 101(c) of the 1990 amendments added Section 110(n)(1) to the Clean Air Act:

Any provision of any applicable implementation plan that was approved or promulgated by the Administrator pursuant to this section as in effect before November 15, 1990 shall remain in effect as part of such applicable implementation plan, except to the extent that a revision to such provision is approved or promulgated by the administrator pursuant to this chapter.

42 U.S.C.A. § 7410(n)(1) (West 1995).

45. 136 Cong. Rec. H13104 (daily ed. October 26, 1990).

46. 42 U.S.C.A. § 7506(c)(3)(B)(i) (West 1995). See note 43 supra for full text.

47. The MTC contended that the congressional comments to the legislation indicated that the savings clause applied only to subsection (A) of Section 176(c)(3), governing plans and programs, and was merely intended to clarify that existing conformity assessment provisions were not automatically eliminated by the new law but would remain in effect until the EPA published new conformity regulations. They believed that it had no affect on subsection (B), which provided an alternative basis for project approval for a 12-month period until new rules could be applied. See note 43 supra for full text of section. Under this interpretation, whether the initial conformity assessments were valid was irrelevant. Supplemental Memorandum of Points and Authorities by the MTC in Opposition to Motion for Injunctive and Other Relief (Re: Clean Air Act Amendments of 1990), November 21, 1990, at 14-15.

48. Sierra Club's Supplemental Reply Memorandum Regarding the Effect of the Clean Air Act Amendments of 1990 on its Motion for Injunctive and Other Relief, November 30, 1990, at 20.

49. *CBE v. Deukmejian,* 34 ERC at 1596.

50. Id. at 1599. The court gave particular weight to the statement by Senator Baucus in the Congressional Record addressing the impact of the savings clause:

Of course, under the savings clause such projects must also comply with any conformity requirements in implementation plans that were in effect prior to enactment of these amendments and existing court decisions in interpreting those SIPs.

136 Cong. Rec. S16973 (October 27, 1990).

51. 34 ERC at 1599. According to the court, by its terms the grandfather clause applied only to the 12 months following the date of enactment and therefore would not affect the four projects approved before the amendments were enacted in November 1990. In addition, the projects in a CO nonattainment area would also have had to satisfy the interim requirements in Section 176(c)(3)(B)(ii) to "eliminate or reduce the severity and number of violations of the carbon monoxide standards in the area substantially affected by the project." 42 U.S.C.A. § 7506(c)(3)(B)(ii) (West 1995). See note 43 supra for full text.

52. 34 ERC at 1600. See note 37 supra.

53. Id. at 1598, quoting statements of Senator Baucus, 136 Cong. Rec. S16972 (October 27, 1990).

54. Id. at 1600-1601.

55. Id. at 1602.

56. Order, Case No. C89-2044 TEH (March 11, 1991).

57. The project was the S.R. 237 upgrade to freeway status. See note 34 supra.

58. *CBE v. Deukmejian,* 34 ERC at 1603.

59. "Judge Halts Three Bay Area Road Projects Pending Assessment of Air Quality Impacts," *Environment Reporter,* January 4, 1991, vol. 21, nos. 27-52, 1628.

60. D. Anderluh, "Judge could create freeway roadblock," *San Jose Mercury News,* December 3, 1990.

61. The California Clean Air Act of 1988 (CCAA), Stats. 1988, ch. 1568, which became effective January 1, 1989, mandates that districts prepare plans to meet the generally more stringent state standards. The State of California has established separate standards for some pollutants known as California Ambient Air Quality Standards or CAAQS. Districts designated, pursuant to the CCAA provisions, as having Moderate air pollution are projected to be able to attain the relevant state standard by 1994. Districts with Serious air pollution are projected to be able to meet state standards by December 31, 1997. Districts with Severe air pollution are projected to be unable to meet the state standard until after the end of 1997, if at all.

62. 42 U.S.C.A. § 7511(a) (West 1995). The design value for Moderate HC areas such the Bay Area is from 13.8 to 16.0 parts per hundred million (pphm).

63. 42 U.S.C.A. § 7512(a) (West 1995). The design value for Moderate CO areas is from 9.1 to 16.4 parts per million (ppm).

64. 42 U.S.C.A. § 7506(c)(1)(B) (West 1995).

65. 42 U.S.C.A. § 7506(c)(2)(A) (West 1995).

66. 42 U.S.C.A. § 7506(c)(2)(B) (West 1995).

67. 42 U.S.C.A. § 7506(c)(2)(C) (West 1995). Projects not meeting these criteria may be approved only if "the projected emissions from the project, when considered together with the projected emissions from the conforming transportation plans and programs within the nonattainment area, do not cause such plans and programs to exceed the emission reduction projections and schedules" assigned to them in the SIP. 42 U.S.C.A. § 7506(c)(2)(D) (West 1995).

68. 42 U.S.C.A. § 7506(c)(3)(A)(iii) (West 1995). See note 43 supra for full text.

69. Section 182(b)(1)(A)(i) describes the standards for demonstrating reasonable further progress for Moderate-and-above nonattainment areas:

> By no later than 3 years after November 15, 1990, the State shall submit a revision to the applicable implementation plan to provide for volatile organic compound emission reductions, within 6 years after November 15, 1990, of at least 15 percent from baseline emissions, accounting for any growth in emissions after November 15, 1990. Such plan shall provide for such specific annual reductions in emissions of volatile organic compounds and oxides of nitrogen as necessary to attain the national primary ambient air quality standard for ozone by the attainment date applicable under this chapter. This subparagraph shall not apply in the case of oxides of nitrogen for those areas for which the Administrator determines (when the Administrator approves the plan or plan revision) that additional reductions of oxides of nitrogen would not contribute to attainment.
>
> 42 U.S.C.A. § 7511a(b)(1)(A)(i) (West 1995).

According to the EPA, the 15% rate of progress requirement is intended to be the base program that all Moderate-and-above areas must meet:

> The base program is necessary to ensure actual progress toward attainment in the face of uncertainties inherent with SIP planning, such as emission inventories, modeling and projection of expected control measures. . . .
>
> In those areas where modeling shows that reductions greater than 15 percent are necessary to attain the standard, the area will be required to achieve those additional emission reductions.
>
> General Preamble for the Implementation of Title I of the Clean Air Act, 57 Fed. Reg. 13,499, 13,507 (April 16, 1992).

70. 42 U.S.C.A. § 7511(a)(1) (West 1995).

71. 42 U.S.C.A. § 7512(a)(1) (West 1995).

72. 42 U.S.C.A. § 7506(c)(3)(B) (West 1995). See note 43 supra for full text.

73. 42 U.S.C.A. § 7506(c)(4)(A) (West 1995); see also 136 Cong. Rec. S16974 (October 27, 1990). On June 7, 1991, the EPA and DOT jointly issued interim guidance for determining conformity of transportation plans and projects. The final rule was promulgated in November 1993 and codified at 40 C.F.R. Part 51, Subpart T.

74. The design value for ozone was 14.0 pphm or 14.3% above the federal standard of 12.0 pphm. Therefore, the stricter minimum 15% reduction would apply. The MTC's Letter to Court, dated March 6, 1991.

75. Id. At the time the Bay Area Plan was adopted, the current 8-hour design value for CO was 10.13 ppm. Because the NAAQS for CO is 9 ppm, a 12.6% reduction from the 1990 baseline over the 5-year period would have met the standard. The EPA subsequently revised the design value for the Bay Area to 11.8 ppm or 23.7% above the standard, and the MTC modified its conformity assessment proposal accordingly.

76. The MTC refused to rerun the POLIS land use model to evaluate possible changes between 1991 and 1997 in the regional pattern of land uses, suggesting that the ABAG estimates already took into account the effects of new highway construction in the TIP. Harvey Supp. Decl., supra note 12, at 4-6. In this regard, the Sierra Club contended that the MTC should be required to model population and land use changes for 1997 and also that its projections for the "no-build" scenario should be reduced to account for possible short-term decreases in the overall levels of regional growth if capacity increasing highway projects were not constructed. Plaintiffs also disputed the MTC's contention that the effects of the current TIP were already in the land use forecasts prepared by ABAG for 1997, arguing that POLIS allocated only a given level of growth over the region and that there was no mechanism by which ABAG assumptions about project schedules and effects on growth are compared with existing local land use plans. Even if there were, plaintiffs argued, it would be necessary to "back out" of these projects for purposes of the "no-build" scenario. Plaintiffs' Memorandum in Support of Objections to MTC's Third Conformity Proposal, January 28, 1991, at 19-20. As the defendants noted, however, for the 1997 analysis the "no-build" comparison is irrelevant because, unlike the 2010 assessment, the 1997 "build" scenario is compared directly to the emissions targets. MTC's Reply Memorandum of Points and Authorities Regarding Proposed Conformity Assessment Procedures (1990 Clean Air Act Amendments), February 6, 1991, at 7, n.6.

77. The MTC argued that the 1990 amendments did not require it to adopt any maintenance plans to ensure air quality levels after 1996. It argued that the only mention of maintenance measures in the 1990 amendments was in Section 175A, which provides that a state requesting a redesignation of any nonattainment area that had attained NAAQS for one or more pollutants must submit a revised SIP that provides for maintenance of the NAAQS for at least 10 years after the redesignation. See 42 U.S.C.A. § 7505a(a) (West 1995). But, see the discussion of the EPA's new conformity requirements in Chapter 5.

78. Metropolitan Transportation Commission's Reply Memorandum of Points and Authorities Regarding Proposed Conformity Assessment Procedures (1990 CAA Amendments), February 6, 1991, at 3-4 [hereinafter MTC's Reply Memo (1990 CAAA)]. See 42 U.S.C.A. § 7506(c)(1)(B) (West 1995).

79. Initially, the MTC stated that conformity would be found if projections for the "build" scenario were below the target levels in the Bay Area Plan. Memorandum of Points and Authorities in Support of the MTC's Revised Conformity Assessment Procedures (1990 Clean Air Act Amendments), January 19, 1991, at 23. The MTC later revised this language to reflect the 1990 amendments.

80. The proposed language would have read,

Section 176(c). Conformity to a plan means—

(B) that such activities will not, *considering any growth likely to result from such activities*—

(i) cause or contribute to any new violation of any standard in any area.

The Senate committee report described the italicized provision as requiring "that the impact of adding highway capacity in or near the non-attainment area on patterns of vehicle use in such areas be quantified to the extent feasible." Senate Committee on Environment and Public Works, S. Rep. No. 228, 101st Cong., 1st Sess. (1989).

81. See 40 C.F.R. § 51. 110(a) (1994) (plan must account for increases resulting from "projected growth of . . . motor vehicle traffic").

82. 42 U.S.C.A. § 7506(c)(3)(A)(iii) (West 1995) (see note 43 supra for full text.); see 42 U.S.C.A. § 7511a(b)(1)(A) (West 1995) (the State shall submit a revision of the applicable implementation plan to provide for VOC emission reductions by November 15, 1996 of "at least 15 percent from baseline emissions, *accounting for any growth* in emissions after November 15, 1990" (emphasis added)).

83. Plaintiffs cited the Guadalupe Corridor Project (State Route 87) as one example. The environmental documentation for a large housing development project in the area suggested that the City of San Jose General Plan made the presence of an uncongested highway circulation network a condition precedent to project approval. Plaintiffs alleged that if the highway project was not implemented, between 13,000 and 18,000 housing units might be refused development permission.

84. At the very least, the MTC considered predicting growth likely to result from transportation projects to be "extremely difficult and speculative." MTC's Reply Memo (1990 CAAA), supra note 80, at 12.

85. Id. at 7, n.6. The MTC conceded that omitting this category of growth could have some effect "on the margin" in the 2010 "no-build" analysis by overstating the number of people who might decide not to move to the Bay Area because some highway projects were not built. MTC felt this omission would have an insignificant effect on the overall model. It also pointed out that it would have no impact on the 1997 analysis, because there the "build" scenario is compared to emissions targets and reductions goals, not a hypothetical "no-build" scenario, as in the 2010 assessment. See note 76 supra.

86. Id. at 8, n.7.

87. Declaration of Raymond J. Brady in Regard to the MTC's Proposed Conformity Assessment Procedure, February 26, 1991, at ¶ 7.

88. Affidavit of Dr. Peter R. Stopher Concerning Transportation Projects and Regional Growth, March 6, 1991, at 2 ("It is not necessary that overall levels of regional growth associated with foregoing any particular project be large in order to demonstrate that, at a local level near a hotspot, the project would increase traffic congestion sufficiently to cause an increase in air pollution.")

89. Id. at 5-6.

90. Plaintiffs' Response to MTC's Declarations Concerning Transportation Projects and Regional Growth, March 6, 1991, at 7-8 ("Only by analyzing proposed highway expansion projects in comparison to a 'no-build' baseline that excludes housing and jobs linked to those projects will the process assure that conformity actually occurs.").

91. Section 176(c)(1) provides, in part,

The determination of conformity shall be based on the most recent estimates of emissions, and such estimates shall be determined from the most recent population, employment, travel and congestion estimates as determined by the metropolitan planning organization or other agency authorized to make such estimates.

42 U.S.C.A. § 7506(c)(1) (West 1995).

92. Plaintiffs pointed out that these factors made assumptions about current emissions and projections for the vehicle fleet in future years, based on post-1987 regulations. Plaintiffs' Memorandum in Support of Objections to the MTC's Third Revised Conformity Proposal, January, 28, 1991, at 12.

93. *Citizens for a Better Environment v. Deukmejian,* 34 ERC 1689 (N.D. Cal. 1991).

94. Id. at 1695. As to the proper standard of review, the court rejected *League to Save Lake Tahoe, Inc. v. Trounday* as inapplicable. The facts of that case are given earlier in note 23. The district court concluded that *Trounday* did not require total (or virtually total) deference to the MTC's conformity proposal because it involved a permit decision, whereas here the court was reviewing a procedure ordered to cure a SIP violation. The court explained,

> The *Trounday* court, however, specifically found that no SIP violation had occurred, and in fact expressly distinguished cases involving SIP enforcement. *Trounday,* 598 F.2d at 1174 ("Unlike [other cases] . . . where state officials were admittedly in violation of explicit strategies incorporated in the [SIP], appellees here have each fulfilled their respective obligations under the Nevada Plan").

> 34 ERC at 1694.

In discussing the standard of reasonableness it was adopting, the court went on to declare that

> plaintiffs are not entitled to their every preference, or the "perfect" conformity procedure. They are entitled, however, to a "reasonable" conformity proposal that will achieve the intended purpose of the conformity requirement. Thus, to the extent there are different, reasonable ways of approaching this task, this Court will not substitute its judgment for MTC's. However, if there are substantial flaws that would preclude MTC from determining conformity in a meaningful manner, then MTC will not be in compliance with its conformity obligations, and its choices cannot be said to be reasonable.

> Id. at 1695.

95. Id. at 1691.

96. The EPA has ruled that EMFAC 7f is now the latest emissions model in California and that it must be used for all conformity analyses begun after 1993. 58 Fed. Reg. 62,211 (November 24, 1993).

97. Id. at 1699.

98. 42 U.S.C.A. § 7506(c)(2)(C) (West 1995). See 40 C.F.R. § 51. 410 (1994).

99. 42 U.S.C.A. § 7506(c)(2)(D) (West 1995).

100. MTC's Reply Memo (1990 CAAA), supra note 80, at 14 (citing remarks by the chairman of the Committee on Public Works and Transportation).

101. As the MTC pointed out, placing a project in the TIP does not constitute an approval of the project, but merely allows the sponsor to continue work on the project for it to become eligible for federal funding.

102. MTC's Reply Memo (1990 CAAA), supra note 80, at 18-22. The procedure for reviewing projects for possible delay is the subject of Resolution No. 2107, discussed earlier in the text in connection with the injunction motion.

103. *CBE v. Deukmejian,* 34 ERC at 1696-1697.

104. 42 U.S.C.A. § 7506(c)(3)(B)(ii) (West 1995). The section provides that this determination may be made either as part of the program conformity determination or later, during the environmental review phase of the project. See note 43 supra for full text.

105. *CBE v. Deukmejian,* 34 ERC at 1699.

106. The MTC responded with criteria for identifying projects with significant potential for creating or contributing to CO hot spots as (a) any project located in an area with a 1987-1989 CO concentration level greater than 6.0 ppm, or (b) in a corridor, identified in the MTC's TIP conformity assessment corridor level analysis, where the TIP "build" vehicular CO emissions were projected to be greater than the "no-build" analysis. Identified projects would have to ensure that CO was taken into account during project design and that a rigorous hot spot analysis would be performed at the environmental review stage. The MTC's letter to court, dated June 10, 1991.

107. Order, Case No. C89-2044 TEH (March 11, 1991).

108. Memorandum, Ninth Circuit Court of Appeals. Case No. 91-15151 (October 21, 1991).

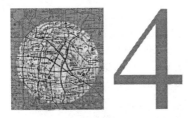

Implementing the Transportation Contingency Plan

The issue of growth was also an important factor in the court's assessment of the adequacy of the Metropolitan Transportation Commission's (MTC's) new contingency measures. In its March 5 Order, the court directed the MTC to both (a) adopt sufficient transportation control measure (TCMs) within 6 months to bring the region back within the reasonable further progress (RFP) line for hydrocarbons (HCs) and carbon monoxide (CO), and (b) adopt, within 45 days, criteria for delaying highway projects with significant adverse impacts on air quality and to apply those criteria within 150 days to the projects in the 1990–94 transportation improvement Program (TIP).[1] At the time of the summary judgment hearing, the MTC had already begun work on a "Draft Implementation Plan for the Contingency Plan in the 1982 Plan" to comply with the Environmental Protection Agency's (EPA's) July 1989 directive to the MTC to implement the Transportation Contingency Plan. The most MTC felt obligated to do, however, was hold a public hearing to solicit comments on projects that might be delayed and suggestions for additional controls, to analyze additional controls, and to hold a final public hearing before adopting any specific measures.

Plaintiffs objected to the Draft Implementation Plan's lack of commitment to specific reductions in emissions, lack of a schedule for the reductions that might be obtained by new TCMs, and lack of commitment to produce attainment as soon as possible. They also complained about the MTC's proposal to exempt certain categories of projects in the TIP from further review for delay without considering either their air quality impacts or their effect on vehicle trips (VTs) and vehicle miles traveled (VMT). In addition to the plaintiffs, the EPA also reviewed and commented on successive drafts of the proposed contingency measures, suggesting changes to comply with the Clean Air Act. The MTC conducted public hearings and considered the comments.

These efforts eventually resulted in the adoption of two resolutions by the board of the MTC. Resolution No. 2107, previously discussed, established criteria for delaying projects until they could be evaluated in the context of an overall air quality plan revision and new attainment demonstration. Resolution No. 2131 contained a number of new TCMs designed to reduce traffic and encourage alternative travel modes. Plaintiffs challenged the adequacy of these new control measures, particularly the failure to consider growth in emissions since 1987. The court ultimately held that the MTC would have to account for some additional growth occurring after the deadline had passed, although not as much as the plaintiffs wanted.

Delaying Highway Projects

In its Draft Implementation Plan, the MTC agreed it would review highway projects with a potentially significant adverse impact on air quality; however, it proposed to exempt any projects that were primarily for safety or that were needed to increase capacity on roadways. Although the MTC agreed that postponing new constructions might cause a shift to alternative transportation modes, such as transit or carpooling and vanpooling, it cautioned that delaying highway projects could have serious consequences for the region, including:

- Increasing freeway and arterial congestion, due to the lack of viable alternative modes for many commuters and growing truck and commercial vehicle use of existing facilities

- Shifting development and the resulting traffic to other locations in the region, and other freeway facilities
- Shifting population and economic growth outside the region.

Delaying projects could also have economic impacts due to cost escalation. For these reasons, the MTC concluded, any such proposals deserved "careful review and a thoughtful response, to assure that significant adverse impacts on air quality will actually be reduced if delays are implemented."[2]

The EPA, however, objected to the MTC's proposal to exempt all capacity-increasing projects as inconsistent with the purpose of the Contingency Plan, because it gives priority to roadways over clean air. EPA insisted that the proper place to balance air quality and transportation needs was in the context of the upcoming revision to the State Implementation Plan (SIP). Federal regulators also warned the MTC that any increased emissions from new transportation projects would have to be offset by reductions from stationary and other sources. Absent a new plan that addressed all factors affecting air quality, the EPA asserted that all capacity-increasing projects would have to be delayed pending the approval of a new SIP.[3]

Resolution No. 2107, as finally adopted, did subject capacity-increasing projects to possible delay but only if the final environmental documentation demonstrated that the project would significantly impact air quality and no adequate mitigation measures were available to neutralize the negative impacts. The MTC developed a four-step process to assess the potential air quality impacts of projects in the TIP, to identify those that would need to be considered for delay in the Contingency Plan:

Step 1. First, the potential impacts of general categories of highway project types was determined, based on the program codes assigned by the state Department of Transportation.

Step 2. Next, the TIP projects in categories judged to have potential impacts in Step 1 were assessed individually, to determine the possible impact of each of the project segments.

Step 3. Subsequently, where certified environmental documentation was available, this analysis was used to determine the air quality impacts of the entire project, to confirm the preliminary determinations assigned to the project segments in Step 2.

Step 4. The final step in the air quality analysis was to determine which projects needed further review in the Contingency Plan.[4]

In Step 1, projects having potential impacts on air quality, such as facility additions and capacity enhancements, would be initially segregated from those, such as rehabilitation work and some operational improvements, that would not.[5] Only projects with completed and approved environmental documentation would be evaluated in Step 2. Those that implemented adopted TCMs, such as high-occupancy vehicle (HOV) lanes, park-and-ride facilities, or carpool/vanpool and transit incentives, would be considered potentially beneficial (B), but those that would not significantly affect travel patterns would receive a neutral (N) designation. Any projects that would increase the capacity of a transportation facility or provide access to undeveloped or underdeveloped areas would be classified as potentially detrimental (D). Sponsors of projects designated "D" were advised to pay special attention to ensure that TCMs were addressed both in the environmental evaluations and in the project design.

In Step 3, those projects given preliminary beneficial or neutral designations in Step 2 would be subject to a detailed analysis, based on the completed environmental report. If the certified environmental documents showed the project would improve air quality or not have a significant impact, or that mitigation measures could neutralize any negative impacts of the project, it would not be subject to further delay. Any project identified as having a significant adverse impact on air quality, which did not include the required mitigation measures from the environmental document, would be classified "D." The MTC defined a *significant adverse impact* as one that produced a net increase in HC emissions between the "build" and "no-build" conditions,[6] or one that caused CO emisions to exceed federal or state emissions standards in the vicinity of the project, or both.[7]

In Step 4, projects designated as potentially detrimental to air quality in the preliminary evaluation in Step 2, or lacking required mitigation measures in Step 3, would have to undergo a further review in the Contingency Plan; however, any projects exempt from the Clean Air Act's conformity requirements would not be subject to any potential delays.[8] If the completed air quality analysis showed that the project would have a significant negative impact on air quality, and inadequate or no mitigation was recommended, or the project sponsor had not committed to carrying out the recommended TCMs, the project would be placed on the list of those to be delayed until the MTC and the sponsor agency agreed on design changes or other mitigation measures; otherwise, the project would be allowed to proceed.

Projects without environmental documentation would remain candidates for delay until the document was completed and certified. Meanwhile, the MTC would work with the project sponsors to review their project designs and ensure that the TCMs to mitigate air quality impacts were adequately considered prior to completing the environmental analysis. Acceptable TCMs would include HOV lanes, Traffic Operations Systems (including ramp metering, incident management, and message signs/traffic advisories), park-and-ride lots, freeway bus turnouts, transit right-of-way reservation, and traffic mitigation requirements for new development near freeways.

MTC agreed to apply these criteria until a new SIP for the Bay Area was approved by EPA. The plaintiffs did not further challenge the MTC's procedures for delaying highway projects, and the court did not rule on their adequacy. They did, however, challenge the sufficiency of the agency's response to the court's order to adopt additional TCMs to remedy RFP shortfalls.

Remedying RFP Shortfalls

Although the court had ruled that conformity assessments did not have to consider additional net growth from new freeway construction, the Sierra Club pressed its contention that the agency's policies were making air pollution worse because of more cars on the road. Plaintiffs insisted that the MTC had to somehow account for higher pollution levels in the region stemming from the increasing traffic. In their view, the Transportation Contingency Plan obligated the MTC to limit overall emissions levels consistent with achieving the National Ambient Air Quality Standards (NAAQS) for both ozone and CO. The agency, on the other hand, believed it had only to reduce emissions by a set amount, as stated in the Bay Area Plan, and not to compensate for any additional pollution caused by unexpected traffic growth. Their dispute revisited some of the same ground covered in the conformity assessment. Plaintiffs charged that the contingency provisions should be broadly interpreted as imposing specific, enforceable obligations on the government agencies to reduce pollution. The MTC countered that these provisions were intended to be more narrow in scope and that past administrative practices and EPA regulations should govern their interpretation.

The question of how far the MTC had to go to meet its responsibilities under the Contingency Plan came down, in part, to whether the agency could measure its compliance based on the assumptions and methodologies in the original Plan or had to take account of new information from the 1990 inventory. That data, prepared for use in the conformity assessment, suggested that pollution levels had not fallen as predicted but had in fact increased. Once again, the enactment of the 1990 amendments figured prominently in the debate, with the Sierra Club taking the position that although the amendments might have imposed additional requirements, they did not vitiate existing SIP provisions, and the MTC asserting that they established a whole new framework that overrode any prior commitments. The whole issue again put the MTC in the uncomfortable position of defending its efforts to reduce traffic, rather than its more accustomed role of building highways to meet travel demand.

In its administrative order, the EPA had directed the MTC to quantify the reductions that new TCMs would have to achieve and the time frame in which to accomplish them. It ordered the agency to pay particular attention to measures that would begin to reduce emissions immediately or in the near future, and to emphasize alternatives to single-occupant vehicle (SOV) trips rather than just reducing congestion or increasing speeds.[9] The draft Implementation Plan included a preliminary screening list of more than 100 potential measures to further reduce emissions and established criteria for reviewing them.[10] Candidate TCMs were evaluated for (a) their effectiveness in addressing the identified RFP shortfalls, (b) whether they could be implemented promptly, (c) whether they could be funded, and (d) whether they could be implemented solely by MTC or needed the approval of other authorities. The EPA, already reflecting the thinking going into the proposed Clean Air Act amendments, was insisting that the transportation planners would have to do more.

The EPA concluded that the measures MTC proposed were simply not adequate. In particular, it found that the MTC had failed to ensure that emissions reductions for both CO and HC, had failed to address shortfalls for the transportation sector, and had failed to satisfy SIP approval criteria, including identifying resources committed to ensuring implementation of the measures as expeditiously as practicable. As for the shortfalls, the EPA took the position that measures to correct the deficiencies must account for any emissions increases due to growth in vehicle travel above that anticipated in the Bay Area Plan. The MTC, however, proposed only to remedy the difference between the emissions reductions expected from the inspection and maintenance (I/M) program and the 10 TCMs in the Plan and the reductions actually obtained from those measures. The EPA concluded that this definition of RFP was incomplete because it ignored the effect of growth in vehicle travel. The EPA advised the MTC that its contingency measures must be sufficient to reduce current emissions levels in the transportation sector "to the *emission levels for 1987* committed to in the 1982 Plan."[11] The MTC also had to justify its decision not to adopt any Section 108(f) TCMs by showing that they would result in "substantial and long-term adverse impacts" and that the standards could be attained without them.

The "2131" TCMs

In its March 5th Order, the district court had ruled that the MTC's efforts to comply with the EPA's order were "too little" and came "far too late" to avoid liability for failure to implement the Contingency Plan.[12] It gave the MTC 6 months to adopt sufficient additional TCMs to bring the region back within the RFP line for both HC and CO. The MTC responded by approving Resolution No. 2131, containing 16 transportation control measures designed to reduce automobile use (the "2131" TCMs). Also included in the resolution were some measures that had previously been adopted and implemented but were not part of the original Bay Area Plan. As shown by Table 4.1, which lists the projected annual and cumulative reductions for both HC and CO expected from the "2131" TCMs, the MTC asserted these new control measures would reduce average HC emissions over a 7-year period by an additional 3.83 tons per day (tpd) and CO emissions by 74.1 tpd.

The "2131" TCMs fell into five different categories: (a) market-based strategies, (b) enrichment of commuter options, (c) congestion management, (d) information and education, and (e) local TSM programs. The MTC rejected other potential TCMs either because funding was unavailable or unlikely in the near future or because they were deemed to be "manifestly lacking" in political support.[13]

Plaintiffs maintained, however, that the estimated reductions for both HC and CO were wholly inadequate to make up the existing shortfalls, and therefore the MTC had failed to comply with the court's order. The Sierra Club also criticized the new control measures as vague and for needing action by other agencies or the state legislature to be implemented. It also charged

TABLE 4.1 Annual Emissions Reductions From "2131" TCMs

Year	1990	1991	1992	1993	1994	1995	1996	Total
HC	1.21	0.54	0.54	0.74	0.58	0.13	0.09	3.83
Cumulative		1.75	2.29	3.03	3.61	3.74	3.83	
CO	19.12	10.05	14.44	15.78	10.76	2.25	1.60	74.1
Cumulative		29.2	43.6	59.4	70.2	72.4	74.1	

that the MTC had failed to adopt all "presumptively" available Section 108(f) TCMs, but for political reasons had instead chosen only certain measures and only particular levels of control, rejecting alternatives offering greater reductions and earlier progress toward attainment. As the plaintiffs also pointed out, many of the anticipated benefits from the proposed TCMs would not be felt for several years. The 16 adopted TCMs and their expected individual contributions to HC and CO emissions reductions, when fully implemented, are shown in Table 4.2, including the name of the agency responsible for implementing them. Emissions reductions were derived by multiplying the anticipated base emissions in the year 1997 by assumed percentage reduction factors for each TCM. These numbers were estimated using the EMFAC 7d emissions factors, an earlier version of the EMFAC 7e factors discussed in the preceding chapter. The totals for all 16 TCMs match the figures shown in Table 4.1.

The Adequacy
of the Contingency Measures

In January 1991, the Sierra Club brought a motion for contempt against the MTC, joined in by Citizens for a Better Environment (CBE), alleging the agency had failed to adopt sufficient contingency TCMs to comply with the court's March 5 Order.[14] The main issues centered on (a) how to define RFP, given that the 1987 deadline had already passed, (b) the appropriate assumptions and methodologies to calculate emissions reductions from the "2131" TCMs and the other control measures, particularly the I/M program,

TABLE 4.2 Estimated Emissions Reductions From "2131" TCMs

"2131" TCMs Description		Emissions Reductions (tons/day)		*Responsible Legal Authority*
		HC	CO	
No. 13.	All bridge tolls to $1	−.19	−3.2	Legislature
No. 14.	Bay Bridge toll to $2	−.15	−2.6	Legislature
No. 15.	Increase gas tax 9¢ per gallon	−.57	−10.3	Legislature and voter approval
No. 16.	New rail starts agreement	−.08	−1.4	Transit Authority
No. 17a.	Continue post-earthquake ferry service	−.02	−.36	Local cities and Transit Authority
No. 17b.	Continue expanded BART service	−.25	−4.4	Bay Area Transit Authority
No. 18.	AMTRAK service to Sacramento	−.07	−1.1	State
No. 19.	Upgrade CALTRAIN service	−.11	−1.9	CalTrans
No. 20.	Regional HOV system plan	−.25	−4.5	CalTrans
No. 21.	Regional transit coordination	−.05	−0.8	Transit operators
No. 22.	Voluntary employer transit subsidies	−.06	−1.0	Local employers
No. 23.	Employer audits	−.16	−2.9	RIDES
No. 24/25.	Expand arterial signal timing program and renew previous signal timing programs	−1.42	−26.5	Local cities, CalTrans
No. 26.	Freeway incident management	−.36	−11.6	CalTrans, CHP
No. 27/28.	Update MTC guidance on local TSM programs	−.09	−1.5	MTC, cities, and counties
Total		−3.83	−74.1	

and (c) whether the MTC had to account for unanticipated growth in vehicular traffic. Also, the parties again disputed the effect the 1990 amendments had on the commitments made in the Bay Area Plan. The MTC insisted that the 1987 standards in the Plan were no longer relevant because the amendments had redefined the standards for RFP. Once again, the Sierra Club responded that the savings clause in Section 110(n) preserved the existing RFP commitments until the EPA approved a new SIP.

Bay Area Plan Ozone RFP Standards

Figure 4.1 presents the relationship between the projected levels of HCs produced by all sources for the years 1979 through 1993, and the RFP standard for achieving reasonable further progress toward meeting the NAAQS. The upper curved line with the shaded box markers represents the baseline HC emissions, that is, without any additional control measures other than those programs already in place before 1982. Values are expressed in tons per day (tpd). These data were derived from the MTC's initial projections for uncontrolled HC emissions estimates in the Bay Area for the years 1979, 1987, and 2000, shown in Figure 2.1. Estimates for the interim years in the figure were interpolated based on the presumed effectiveness of the pre-1982 control measures.

As shown by Figure 4.1, in 1979, estimated HC emissions were 732 tpd. The MTC anticipated they would decline to 515 tpd by 1987 due to existing control measures, as indicated by the initial steep downward slope of the line. After 1987, average daily HC emissions were expected to gradually increase without adoption of the Bay Area Plan's controls. The solid horizontal line represents the RFP standard for HC emissions established in the Bay Area Plan, in this case 430 tpd.

Moving to the right, along the RFP line, one can see how much emissions must be reduced each year from the projected baseline to meet the 430 tpd target. In accordance with EPA regulations, the attainment demonstration in the Bay Area Plan established that, beginning in 1983, the control measures listed in the Plan would produce a straight-line reduction in total expected HC emissions from baseline levels, as indicated by the line in the figure with solid box markers (the 1982 Plan Inventory).[15] The vertical distance between the two lines measures the emissions reductions needed to attain RFP. For the region to meet RFP by 1987, total HC emissions had to fall an additional

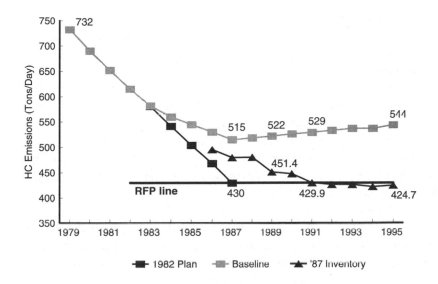

Figure 4.1. Bay Area Plan RFP Standards for Regional HC Emissions, 1979–1995

85 tpd below the baseline HC levels by that year (515 minus 430). According to the Plan, this would be achieved with a 56 tpd reduction coming from controls on stationary sources and the remaining 29 tpd from the vehicle I/M program, assuming a 25% effectiveness rate.

Under the Contingency Plan, the MTC had agreed to make up any short-falls in RFP.[16] Because the Bay Area Plan had established RFP in terms of the stated emissions reductions and overall emissions levels, the court had ruled these commitments were enforceable SIP provisions. In its contempt motion, the Sierra Club argued that the court had directed the defendants to achieve both an 85 tpd reduction in HC emissions and the target level of 430 tpd in total residual HC emissions. More important, though, plaintiffs pointed out that the baseline emissions had not remained fixed. As shown in Figure 4.1, the growth in vehicular travel between 1987 and 1989 produced an extra 7 tpd in baseline emissions (from 515 to 522 tpd), indicating that now a 92 tpd reduction would be required to attain RFP by the end of 1989, instead of the original 85 tpd.

Although the state legislature had strengthened the I/M program, revised estimates showed that even this enhanced smog check program was only

18% effective (instead of 25%). As of 1989, it was producing a mere 22 tpd reduction of HCs.[17] Estimated HC reductions from all stationary source control measures produced just 48.6 tpd in reductions,[18] still 7.4 tons short of the 56 tpd target in the Plan. Thus, as of the court's initial ruling, total HC emissions reductions of only 70.6 tpd were being achieved, which when subtracted from the 1989 baseline figure (522 minus 70.6), left 451.4 tons in residual emissions, as indicated in Figure 4.1. This was 14.4 tpd less than the required 85 tpd reduction and more than 21 tons short of the revised 92 tpd figure. Therefore, the court concluded, whether measuring by the 85 tpd reduction target or the residual 430 tpd emissions level, the defendants had failed to achieve RFP for HCs.[19]

The plaintiffs urged the court to find the MTC in contempt, because the measures it had adopted to comply with the court's remedial order accounted for only 3.83 tpd, not nearly enough to erase the shortfall. Even worse, projections from the 1990 inventory showed that actual regional ozone levels were higher than expected, which meant even greater reductions would be needed to make RFP. They also contended that the agency could not show RFP had been met for CO emissions on either the regional or the subregional level. In response, the MTC argued that the issue was no longer as simple as the plaintiffs made it appear. By enacting the 1990 amendments, the agency explained, Congress had effectively redrawn the RFP lines for HC and CO emissions; because these new provisions took effect immediately, they should be used in evaluating RFP instead of the old lines. Because of this, the MTC insisted that it was "no longer constrained" by the "assumptions and methodologies" in the Bay Area Plan and filed a countermotion for partial summary judgment, claiming that the 1990 amendments had eliminated their obligation to adopt any additional TCMs.

Judge Henderson ruled on the contempt motion on August 19, 1991, after holding two separate hearings and reviewing numerous written submissions from the parties, including supplemental legal briefs that he had ordered them to prepare. The judge first held that the MTC was still required to comply with the provisions of the Bay Area Plan as well as any interim regulations contained in the 1990 amendments.[20] The amendments did not, the court concluded, erase the defendants' obligations under the 1977 Clean Air Act and the Bay Area Plan. Rather, Congress had intended to hold local agencies to their existing commitments pending a formal SIP revision. The court explained that,

> Such an interpretation best reconciles and harmonizes the purposes of the Clean Air Act and the provisions of the 1990 Amendments. It is most consistent with the overall purpose of achieving healthy air as expeditiously as practicable and the statutory scheme of retaining the enforceability of existing SIPs until replaced.[21]

The court noted that the amendments did not change the basic concept of RFP as the annual incremental reductions necessary to achieve NAAQS.[22] The Bay Area Plan reflected this and committed the MTC to achieving the "RFP lines" contained there. The 1990 amendments merely included additional requirements that had to be met during the transitional period and incorporated into the new SIP.

The issue was not, explained the court, an either-or proposition—whether RFP obligations were governed only by the original SIP or only by the 1990 amendments during the period before a new SIP was produced. The reductions required by the 1990 amendments clearly applied during the interim period in the sense that the defendants would have to satisfy the required reductions by 1995 for CO and by 1996 for ozone; however, that particular issue was not before the court. The real question was whether the original RFP obligations in the Bay Area Plan remained enforceable or if RFP was to be governed solely by the 1990 amendments, a question not specifically dealt with in the legislation.

Reviewing the legislative history and purpose of the 1990 amendments, the court concluded that nothing in the amendments suggested any congressional intent to abolish the commitments in existing SIPs. Indeed, the savings clause in Section 110(n) specifically provided that "any provision" of an existing SIP would remain in effect until the EPA approved a revision.[23] Rejecting the MTC's suggestion that this created a conflict between the agency's commitment to satisfy the RFP benchmarks in the Bay Area Plan and its obligations to demonstrate the reductions required by the 1990 amendments, the court reasoned that Congress intended the new RFP standards to operate only as "outside limits" during the SIP revision process. If the existing RFP commitments required less action than the new RFP schedules, there would be no conflict. On the other hand, if the existing commitments required additional steps beyond what was required to satisfy the new deadlines, this would still be consistent with the congressional purpose of attaining the NAAQS "expeditiously."[24]

Still, the parties disagreed as to how the emissions level component of RFP in the Bay Area Plan should be evaluated. MTC contended that current emissions levels could be assumed, based on subtracting estimated emission reductions from the Plan's projected uncontrolled emissions levels. Plaintiffs, on the other hand, maintained that undertaking a separate analysis or modeling effort to assess current emissions levels would be more consistent with the purposes of evaluating RFP. For the court, the real issue was not so much the conflict in modeling approaches but the fact that MTC had committed itself to adopting adequate contingency measures in the event RFP was not met by the statutory deadline. Although the court would not become involved in deciding what would be needed to achieve NAAQS under the new planning process required by the 1990 amendments, it could and would enforce specific commitments made in the Bay Area Plan. That still left open the question of whether to judge those commitments based on the original assumptions and methodologies in the Bay Area Plan or by reference to the revised air quality estimates contained in the 1990 inventory. Plaintiffs believed that the new estimates more accurately reflected actual pollution levels and should be the basis for determining liability. Here, however, the court would side with the position taken by the MTC, holding that the agency could rely on its original baseline estimates to calculate emissions reductions expected from the various control programs.

The 1987 Emissions Inventory

Notwithstanding the amendments, the MTC maintained that it could satisfy RFP for both HC and CO, even without the "2131" TCMs, basing its contention on newly revised estimates of the effectiveness of both the stationary source and the transportation controls. To begin with, the changes the state legislature had made in the I/M program were expected to increase its effectiveness to 28% over the next several years. MTC insisted that improved performance from the stationary source controls, the new enhanced I/M program, and other recently adopted measures would achieve the necessary reductions in HC and CO emissions by *1991*. Using the revised estimates of I/M program effectiveness, the new figures (the 1987 inventory) were derived by subtracting the expected reductions from the original baseline emissions projections in the Bay Area Plan. Again, the Plan had assumed 1987 emissions levels would be 515 tpd for all sources of HCs, and 1,934

TABLE 4.3 1987 Inventory of Hydrocarbon Emissions (Tons/Day)

Year	1989	1990	1991	...	1995
Baseline	522	526	529		544
Stationary source controls					
Adopted controls	−44.9	−49.5	−63.9		−69.5
Court mandates	−3.7	−3.7	−4.7		−12.2
Transportation controls					
I/M program	−22.0	−25.0	−30.5		−37.6
Subtotal	−70.6	−78.2	−99.1		−119.3
1987 inventory	451.4	447.8	429.9		424.7

tpd for CO from the transportation sector, as depicted in Figures 2.1 and 2.2. Although RFP could no longer be established as quickly as before, MTC was confident it could be achieved in the near future.

Hydrocarbon RFP

The MTC conceded that uncontrolled emissions would continue to increase, and that baseline emissions would reach 529 tpd by the end of 1991, as shown in Figure 4.1. This meant it would be necessary to achieve a target reduction of 99 tpd (compared to the 85 tpd called for in the Plan) to reach the 430 tpd RFP level. Although the agency asserted that it was at most responsible only for the 85 tpd reduction, even assuming it had to account for the modest increase, the MTC insisted that its 1987 inventory demonstrated that sufficient reductions could be attained by that date to offset the projected upturn. Thus, RFP could be achieved by then, even as defined by the plaintiffs.

Table 4.3 shows how the MTC calculated the 1987 inventory for HC emissions. Baseline emissions for 1989 through 1995 are listed along the top row, with the expected reductions from the various programs underneath. For 1989, baseline emissions are 522 tpd. Combined reductions in HC emissions amounted to about 71 tpd, as described earlier. If this amount is subtracted from the baseline, the result is 451.4 tpd, which still exceeds RFP, reflecting the court's previous calculations.

To reach the RFP standard for ozone, the MTC took credit for projected improvements both in stationary source controls and from controls in the transportation sector. The MTC predicted that by year-end 1991, the originally adopted stationary source controls (63.9 tpd), together with the four additional measures mandated by the court, covering reciprocating engines, solvents, and pesticides (4.7 tpd), would contribute nearly 69 tpd in reductions, well above the original 56 tpd estimate. Added to reductions from the improved I/M program operating at 24% effectiveness (30.5 tpd), the combined pollution control strategies would produce a total of 99.1 tpd in HC reductions. Subtracting these from the 1991 baseline would result in an emissions inventory level of 429.9 tpd (529 minus 99.1), fractionally below the RFP limit. When added to the other reductions listed in Table 4.3, the 3.83 tpd from the "2131" TCMs provided a small margin of error.[25]

The relationship between the revised 1987 emissions inventory, the baseline emissions, and the RFP line is illustrated graphically in Figure 4.1. The 1987 inventory (from Table 4.3) is shown by the line with the triangle markers. This line just crosses the RFP line in 1991, indicating that RFP would be attained by that date. As shown in the table, the MTC also projected that it could maintain RFP after 1991, because overall emissions would remain below the 430 tpd RFP line through 1995, even allowing for expected growth. Adding the projected annual reductions from the "2131" TCMs for the years 1990 through 1995, shown in Table 4.1, would bring this line down even lower. To summarize, although baseline HC emissions (again without the Bay Area Plan controls) would rise to 529 tpd by 1991, according to the MTC, a series of annual reductions would gradually reduce inventory levels below the RFP target of 430 tpd by that time.

Regional CO RFP

Although the court had not endorsed a particular RFP standard for CO in its summary judgment ruling, the Sierra Club argued that the MTC was responsible for attaining RFP for CO on both a regional and subregional level. The plaintiffs proposed that there were two regional CO RFP standards in the Bay Area Plan, one associated with vehicular emissions and a second, "surrogate" standard related to vehicle travel. Using the same approach adopted by the court for HC emissions, plaintiffs argued that the MTC had committed to reducing regional CO levels 393 tpd by 1987. Because under

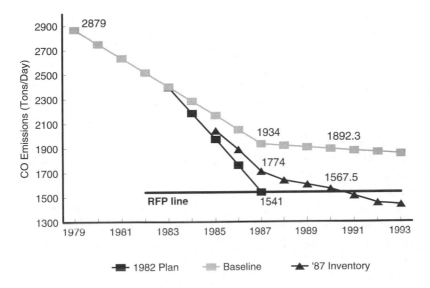

Figure 4.2. Regional Transportation Sector CO Emissions, 1979–1993

the Plan the full CO reduction was to come from the transportation sector, which was expected to contribute 1,934 tpd to the 1987 baseline (see Figure 2.2), plaintiffs reasoned that RFP for the transportation sector alone would be 1,541 tpd (1,934 minus 393).

Figure 4.2 shows the projected CO emissions from the transportation sector for the years 1979 through 1993, derived from the 1979 CO inventory in Chapter 2. As in Figure 4.1, the shaded box markers again represent the baseline emissions and the solid box markers the expected emissions inventory after applying the planned controls. Mobile source emissions accounted for 2,879 tpd in 1979. Here again, improvements in engine technology were expected to gradually reduce that figure. Most of the additional planned reductions were to come from the I/M program. Of the 1,934 tpd in mobile source emissions projected for 1987, 1,468 tons were expected to come from cars, light trucks, and medium duty vehicles that were subject to inspection.[26] According to the Bay Area Plan, it was initially expected that the statewide smog check program would reduce emissions from these sources by 25%, or 367 tpd. The Plan's control strategy had consisted of this program plus an additional 26 tpd reduction from the original 10 TCMs. Together, these controls would have reduced CO emissions down to the deduced RFP line by 1987, as shown in Figure 4.2. However, as with ozone, the projected CO reductions were not met. The original 10 TCMs resulted in only 16 to 20 tpd in CO reductions through 1987, rather than the expected 26 tpd.[27] But the primary reason that RFP was not met for CO was again the disappointing performance of the I/M program, which was reducing CO emissions by only 144 tpd at the start of 1987.[28] As indicated by the line with triangle markers in the figure, total reductions for 1987 amounted to only about 160 to 164 tpd, leaving a shortfall of more than 230 tons.

As it had for HC emissions, the Sierra Club again insisted that the projected 74.1 tpd in CO reductions from the "2131" TCMs was clearly inadequate to establish RFP, even on top of the estimated 160 to 164 tons in reductions from existing controls. The plaintiffs pointed out that MTC's own latest estimates of actual emissions from the transportation sector ranged anywhere from 1,860 to 1,937 tpd for the year 1986, representing a 319 to 396 ton shortfall from the 1,541 tpd target. This was a clear indication that despite the adopted controls, pollution was getting worse.[29]

The MTC insisted, though, that the Sierra Club's 1,541 tpd RFP standard, or any regional CO standard, was flawed. They maintained that no regional RFP standards for CO were included in the Bay Area Plan, as was done for HCs, because CO pollution is primarily a local rather than a regional problem. Instead, the CO control strategy was primarily aimed at eliminating hot spots not achieving any specified reductions in CO levels, and therefore the regional CO emissions figures in the Plan were purely informational. Because it had carried out all of the contingency measures for CO, including the Guadalupe Light Rail Transit, the Synchronized Signal Control Program, and a TSM plan and parking policies for downtown San Jose, the MTC insisted that it had made reasonable progress in reducing CO emissions.

In any event, the MTC argued that the downward trend in CO emissions, due to existing programs and improvement in vehicle design, would contribute to reaching the Plan's goals by 1991. Using the assumptions in the Bay Area Plan, uncontrolled CO emissions from the transportation sector would decline from 1,934 tpd in 1987 to 1,892.3 tpd by 1990, the last full year prior to the plaintiffs' motion. With existing controls, emissions would be reduced to 1,567.5 tpd (see Figure 4.2). Again, according to the MTC, the adopted control programs would bring actual emissions down to less than 1,541 tpd by 1991.

Table 4.4 shows how the 1987 inventory for CO was reached. It illustrates that, by year-end 1990, the enhanced I/M program would provide 306.8 tpd

TABLE 4.4 1987 Inventory of Regional Carbon Monoxide Emissions (Tons/Day)

Year	1989	1990	1991	1992
Baseline	1,907.2	1,892.3	1,878.4	1,880.5
I/M program	−284.8	−306.8	−349.4	−394.9
10 TCMs	−18	−18	−18	−18
Subtotal	−302.8	−324.8	−367.4	−412.9
1987 inventory	1,604.4	1,567.5	1,511.0	1,457.6

in reductions (now assuming a more modest 20.9% effectiveness rate rather than the initial 25% estimate).[30] The original 10 TCMs would add an average of about 18 tpd, for a total reduction of 324.8 tpd. Subtracting this from the 1990 baseline figure would leave total CO emissions just 26.5 tpd over the plaintiffs' assumed RFP standard. The MTC contended that improvements in tailpipe emissions would make up any difference by 1991, even before any of the "2131" TCMs were to take effect. By 1991, the I/M program was expected to eliminate close to 24% of CO emissions. Applying this revised effectiveness rate to the base mobile source emissions, the smog check program would generate 349.4 tpd in emissions savings, which when combined with the reductions from the 10 TCMs, would be enough to demonstrate compliance for the region.

The 1987 inventory data from Table 4.4 are also illustrated in Figure 4.2, which depicts the revised CO emissions inventory dropping below the horizontal RFP standard between 1990 and 1991. Again, by taking into account the additional yearly reductions from the "2131" TCMs (see Table 4.1), the region would meet the standard even sooner.

"Surrogate" CO RFP Lines (VTs and VMT)

The anticipated amount of vehicle travel for 1987, based on projected population and employment, was an important element in determining the emissions levels used in developing the Bay Area Plan. Despite the MTC's demonstration regarding regional CO emissions, the Sierra Club maintained MTC had failed to achieve the overall reductions in VTs and VMT projected in the Plan. As discussed in Chapter 3, the plaintiffs contended that the actual traffic levels in the region were exceeding the Plan's initial estimates. The Plan had proposed establishing a system to monitor trends in regional travel to compare the actual growth in travel with the original assumptions. Estimated noncommercial VTs in 1979 were 10,000,000, increasing to 11,250,000 by 1987, and daily VMT were expected to jump from 77,300,000 to 90,400,000 miles during this period. The adopted transportation controls were intended to reduce vehicular travel from these baseline projections.[31] Plaintiffs viewed these VT and VMT estimates as "surrogate" measures for RFP, based on the fact that the Transportation Contingency Plan stated that the annual RFP reports would assess "the growth in vehicle travel in the region" (see Appendix B).

Taking into account the expected reductions from the Bay Area Plan, the Sierra Club estimated the "surrogate" RFP figure for 1987 at 89,189,436 vehicle miles. The observed 1987 daily VMT level was 105,700,000, well above even the baseline level. Similarly, the number of all VTs was 15% greater for 1987 than had been assumed for 1980.[32] Plaintiffs insisted that the MTC would have to compensate for this unexpected growth in vehicle use. Indeed, in responding to the proposed contingency TCMs, the EPA itself took the position that the measures were inadequate, because the MTC had ignored the effects of growth in VTs and VMT and that any contingency measures must account for these effects, not just shortfalls in prior TCM implementation. To counter the additional traffic, the MTC would have to deduct from its estimates any growth that occurred between 1987 and 1990.[33]

For its part, the MTC rejected the suggestion that there were "surrogate" regional CO standards in the Bay Area Plan, because the Clean Air Act defined RFP in terms of reducing emissions not VTs or VMT. The MTC reiterated that it did not have to maintain projected traffic levels from the Plan as long as emissions targets could be achieved. As the agency pointed out, even under the 1990 amendments only Serious CO areas have to adopt specific enforceable TCMs to offset any growth in emissions due to higher numbers of VTs or VMT in the area.[34] No such requirements were included for Moderate areas such as the Bay Area. In fact, MPOs in Moderate CO areas are required only to provide forecasts of VMT if the 8-hour CO design value is greater than 12.7 parts per million (ppm), well above the Bay Area level of 10.3 ppm, and even then a SIP revision would not be due until late 1992.[35]

Plaintiffs responded that the Bay Area Plan had specifically included VMT standards and that these should be used to assess conformity regardless of how the 1990 amendments dealt with the issue. They allowed that the required SIP revision could adopt a new trade-off and permit higher vehicle use at the expense of stricter controls on stationary sources if that were deemed appropriate but that, meanwhile, the existing assumptions and methodologies should control. Moreover, plaintiffs argued that the new requirements for Serious or Severe ozone areas were intended to reduce excess emissions in these areas by mandating new TCMs, not as a license for Moderate nonattainment areas to ignore existing control strategies.

Areawide and Hot Spot RFP for CO

The Sierra Club also maintained that the MTC had failed to demonstrate RFP at a subregional level either for the 25 square kilometer area around San Jose or at the designated hot spot. The Bay Area Plan called for reducing the 1987 baseline Areawide CO concentrations by 10,800 g/pk-hr/km^2 from 118,800 to 108,000.[36] Similarly, the Plan projected CO emissions at the First and Santa Clara intersection in San Jose to decline from 8,474 g/pk-hr to just 6,428 g/pk-hr (see Figure 1.3). The RFP report for 1986 showed Areawide emissions to be 115,000 g/pk-hr/km^2, and Hot Spot emissions for the same year were about 7,800 g/pk-hr.[37]

In contrast to its position on the regional RFP standard, the MTC accepted the Areawide and Hot Spot figures as appropriate for measuring RFP, but insisted that they could be met. As it did in its regional HC demonstration, the agency took credit for the projected declines in CO after 1987 to counter poor I/M performance on the subregional level. Using the same methodology as in the Bay Area Plan, the MTC estimated Areawide CO emissions for 1991 at just 95,596 gm/pk-hr/km^2, well below the Areawide RFP standard.[38] The MTC also asserted, based on the same analysis, that even without any additional controls, emissions at First and Santa Clara would have fallen to 5,928.1 g/pk-hr, quite a bit less than the Hot Spot RFP standard, and that the I/M program should further reduce emissions.[39] The agency maintained that in addition to the existing measures, the planned creation of a downtown transit mall would eliminate any potential hot spots. Plaintiffs, however, insisted that, when properly corrected for ambient temperature and average speeds, the Areawide figure was actually 30% above RFP levels, and that given these errors and other mistaken assumptions, the MTC could not prove it was in compliance.

The 1990 Emissions Inventory

The entire RFP issue was further complicated by the initial results coming from the MTC's new travel model program. The Sierra Club contended that the agency was ignoring the effects of regional growth by discussing only the targeted reductions with reference to an out-of-date baseline. The new, updated emissions estimates from the 1990 inventory, developed to meet the conformity requirements in the 1990 amendments, showed that even with all the Bay Area Plan controls in place, the region was in fact well above the RFP targets for mobile sources.

In relying on the 1987 inventory to show compliance, the MTC acknowledged that "the actual vehicular hydrocarbon emissions may be higher than these *assumed* values" but it insisted that, in its earlier summary judgment ruling on the Contingency Plan, the court had fully understood that the 430 tpd target for HC emissions was based solely on the agency's initial baseline projections, and that even if the defendants achieved the stated reductions relative to these estimated uncontrolled emissions, it might still not result in the area's attaining the NAAQS.[40] Plaintiffs responded, though, that the MTC should not be allowed to ignore actual emissions levels in favor of its out-of-date projections.

Unlike the 1987 inventory, which was based on the Plan's baseline forecasts, the 1990 inventory was prepared using the MTCFCAST travel model described in Chapter 3, without any reference to the original HC and CO estimates. Although the MTCFCAST model accounts for most of the traffic in the region, it does not capture all sources of vehicular travel that contribute to pollution. As a result, the 1990 inventory was supplemented with "off-model" trip estimates for trips between home and school, trips by commercial vehicles such as medium and heavy-duty trucks, and interregional trips from and to points outside the Basin. It also incorporated emissions reductions from some of the "2131" TCMs into the model.

The results of the modeling for HC emissions are shown in Figure 4.3, which compares the 1987 and 1990 inventories. Note that the figure reflects only pollutant emissions from the transportation sector. On the basis of this new, updated data, the plaintiffs asserted that even with implementation of

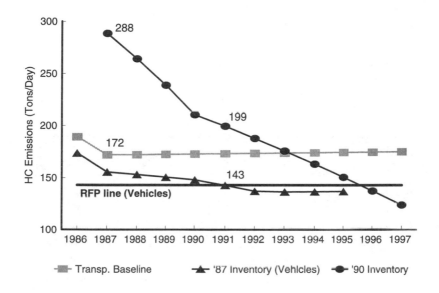

Figure 4.3. 1990 Inventory Transportation Sector HC Emissions, 1986–1997

the "2131" TCMs, the region still would not meet the RFP standards by 1991 for either HC or CO. As stated in Chapter 2, the transportation sector's share of the emission reductions for ozone was 29 tpd. Because the Bay Area Plan projected uncontrolled HC emissions for the transportation sector at 172 tpd, RFP for mobile sources alone would be 143 tpd (see Figure 2.2). The triangle markers in Figure 4.3 represent the vehicular portion of the 1987 inventory, as projected by the MTC. They are derived by subtracting the anticipated reductions from the I/M program from the transportation sector portion of the baseline, shown by the rectangular shaded box markers. The circles represent the revised emissions projections from the 1990 inventory. According to the new inventory figures, actual HC emissions for 1991 from vehicles were 199 tpd, quite a bit higher than the RFP benchmark in the Bay Area Plan—higher even than the initial estimates of uncontrolled vehicle emissions, indicating that ambient pollution had gotten worse rather than better. Plaintiffs interpreted the new estimates to now require the MTC to achieve a total of 56 tpd in HC reductions from the transportation sector (199 minus 143) instead of the original 29 tpd in the Plan.

Comparing the 1987 inventory to the 1990 inventory in Figure 4.2 shows that the later curve appears to have shifted upward from the earlier estimates,

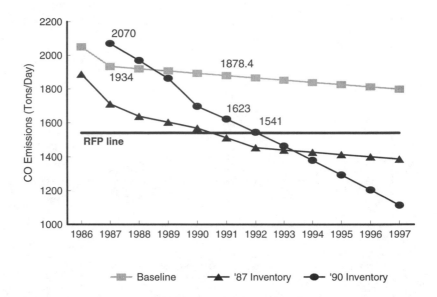

Figure 4.4. 1990 Inventory Regional CO Emissions, 1986–1997

reflecting higher emissions levels in each year through at least 1993. This shift could have resulted from lower than expected emissions reductions from the Bay Area Plan controls, higher than expected emissions caused by unanticipated growth (or else poor performance from the pre-1982 controls), or both. Because in either case the 1987 attainment year figures for the 1990 inventories exceed even the baseline emissions estimates (again represented by the lines with rectangular shaded box markers), it was a clear indication to plaintiffs that, irrespective of any planned emissions reductions, some growth in ambient emissions had taken place over and above that predicted in the Bay Area Plan. That growth must have been due, plaintiffs alleged, to the higher levels of VTs and VMT than originally assumed in the Plan.

The situation was similar for CO emissions. As shown in Figure 4.4, the new CO emissions figure of 1,623 tpd also exceeded the 1,541 tpd RFP benchmark by a wide margin. Again, plaintiffs asserted that the MTC would need to come up with additional CO reductions of 82 tpd to attain RFP.

Plaintiffs took the position that the Contingency Plan required the MTC to account for this increased growth by adopting additional TCMs to bring both regional HC and CO emissions back down to their respective RFP lines.

As depicted in Figures 4.3 and 4.4, without those additional controls this would not occur before 1996 for HC emissions and 1992 for CO emissions.

The MTC responded, however, that the 1990 inventory used an entirely different methodology from the previous assessments in the Bay Area Plan and therefore the two sets of figures were not comparable. For instance, the new, updated data, which would be used to develop the revised SIP, were based on different types and categories of emission sources not included in either the 1979 baselines or the 1987 inventories. Also, the Bay Area Plan had used the EMFAC 6c emissions factors, but the newer projections were based on the recently developed EMFAC 7e factors. In short, the MTC insisted that using the 1990 inventory data to claim that the Bay Area Plan had underestimated emissions levels was like comparing apples and oranges.

Even if the two sets of figures could be compared, the MTC again insisted that its legal obligations would not be affected, because it was not responsible for any increased growth not covered by the original attainment demonstration. To illustrate, recall from Chapter 2 that in the Bay Area Plan, the Bay Area planners used the Livermore Regional Air Quality (LIRAQ) model to compute the 1987 design value of 14.4 tpd for HC emissions in the region. Because this exceeded the 12 parts per hundred million (pphm) federal standard by 17%, the expected emissions in 1987 associated with this design value (515 tpd) would also have to be reduced by 17%, resulting in the 430 tpd RFP target value. Thus, the relevance of the 430 tpd figure as a standard associated with attainment of the NAAQS depended entirely on the accuracy of the projected HC emissions in the Plan, which were based on assumptions about growth trends, energy use, and control programs then in effect.

The MTC's position was basically that it could not be held liable for errors in forecasting. If it turned out that the actual HC emissions were higher than the original baseline projections, the agency argued it would still be responsible only for achieving the same reductions specified in the Bay Area Plan (i.e., 85 tons), but that the residual emissions level would necessarily be different and the court had understood this might not achieve NAAQS. In essence, the MTC contended that it was responsible only for correcting deficiencies in the Plan's pollution controls, not for accounting for unexpected growth. Therefore, it should be held liable, if at all, only for reducing HC emissions 85 tons from the expected 1987 baseline, not achieving an arbitrary 430 tpd target. As the plaintiffs pointed out, however, the court had apparently rejected this view in its summary judgment ruling and had instead held that the MTC would have to meet both standards, in effect deciding that the MTC would be responsible for any additional growth. The controversy therefore came down to whether the agency could measure compliance from the 1987 inventory, which reflected the original baselines, or was accountable for the higher pollution estimates in the 1990 inventory. The MTC maintained that the court's initial ruling was moot in any event, because the 1990 amendments had imposed new RFP standards, which replaced those established in the 1982 Plan and should be used in any evaluation of RFP based on the new figures. Once again, the court had to assess the effect of the 1990 amendments.

The Effect of the
1990 Clean Air Act Amendments

Beyond the issue of how the new emissions estimates were derived, the MTC insisted that the standards in the Bay Area Plan were simply no longer relevant, because passage of the 1990 amendments to the Clean Air Act established brand-new RFP standards based on the new 1990 inventory baselines. The agency maintained that even if the Bay Area Plan remained temporarily in effect pending revision, the 1990 amendments had replaced the original RFP targets with these new standards during this interim period. The 1990 inventory was therefore relevant only to the new standards and should not be measured against the 1987 RFP targets for either HC or CO. This updated analysis, according to the MTC, took into account the additional levels of travel not counted in the 1987 inventory and also additional improvements in the vehicle fleet since 1987.[41]

Hydrocarbons

As noted in Chapter 3, the 1990 amendments provide that by the end of 1996, those Moderate nonattainment areas for HCs, with design values between 13.8 and 16.0 pphm, such as the Bay Area must achieve a minimum 15% reduction from 1990 baseline emissions (accounting for any growth in emissions after 1990). Figure 4.5 shows how the MTC interpreted its obligations under the new law. It compares the results of the 1990 HC inventory with the new RFP standard in the amendments. Projected HC emissions from the 1990 inventory are again shown as solid circles. The estimated vehicular

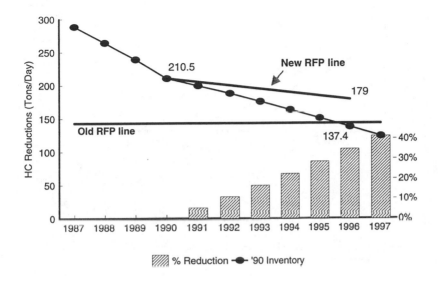

Figure 4.5. Revised RFP Standards for HC Emissions (1990 Amendments)

emissions for 1990 are 210.5 tpd, well above the 143 tpd standard represented by the old RFP line for mobile sources, which is included for reference. A 15% reduction over 6 years would produce a figure of 179 tpd in 1996. The new RFP line can therefore be represented by the downward sloping heavy line drawn between these two points. As long as the projected emissions remain below this line, the region satisfies the interim requirements under the 1990 amendments, which is indeed the case, as shown in the figure. According to the data, by 1996 HC emissions would be 137.4 tpd. This would constitute a 34.7% reduction from the base year, considerably better than the interim standard. As mentioned earlier, the MTCFCAST model took into account the effect of only some of the "2131" TCMs in its projections. By including all the "2131" TCMs, the actual figure for 1996 would drop to 135.1 tpd, reflecting a 35.8% reduction over 1990 levels.

Regional CO

Although the 1990 amendments did not establish any minimum percentage requirements for reducing CO emissions, as for HC, they did adopt

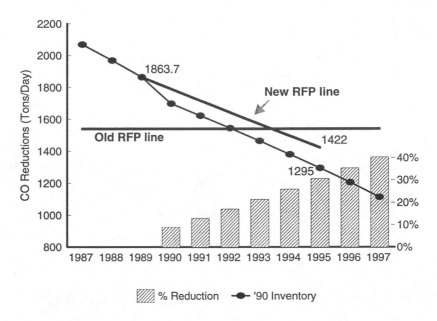

Figure 4.6. Revised Regional RFP Standards for CO Emissions (1990 Amendments)

revised attainment deadlines. Each nonattainment area for CO with a design value of 9.1 to 16.4 ppm was classified as Moderate and given until December 31, 1995, to attain the primary NAAQS.[42] The amendments require a rollback in CO emissions from the appropriate base year equal to the same percentage amount by which the CO design value exceeds the federal 8-hour standard of 9.0 ppm. In March 1991, the EPA advised the MTC that it was adjusting the design value for the Bay Area upward from 10.3 ppm to 11.8 ppm, based on new field measurements. Because the new design value was 23.7% above the federal standard, CO emissions would have to be reduced by that much to show attainment.

Figure 4.6 compares the new CO inventory figures with the adjusted federal standard. According to the 1990 inventory, estimated emissions in 1989, the new base year chosen to reflect the revised design value,[43] were 1,863.7 tpd. A 23.7% reduction would produce an RFP target of 1,422 tpd in CO emissions by 1996, which is actually less than the original 1,541 tpd target. Again, the new RFP schedule is represented by the heavy downward sloping line in the figure. Projected emissions for the target year of 1995

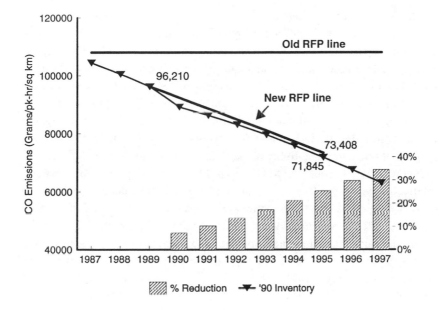

Figure 4.7. Revised Subregional RFP Standards for CO Emissions (1990 Amendments)

were 1,295 tpd, which represents a 30.6% reduction, again exceeding the new RFP standard. This amount drops to 1,275.5 tpd by factoring in all the remaining "2131" TCMs. MTC admitted, however, that four of those TCMs had become stalled for lack of either legislative authority or funding.[44]

Areawide CO

The MTC also asserted that it could meet the new subregional benchmarks. Figure 4.7 compares the new Areawide emissions estimates for the San Jose area with the new federal standard. Using the revised data, Areawide emissions for 1989 were 96,210 g/pk-hr/km^2. They were projected to drop to 71,845 g/pk-hr/km^2 in 1995, a 25.3% reduction, and to 70,824 g/pk-hr/km^2, with implementation of all the "2131" TCMs, a 26.4% reduction.[45] Here again, the emissions reductions exceed the required percentages, and the projected trend line remains below the new RFP line over the relevant period.[46]

The Sierra Club responded to the MTC's argument by insisting it was premature to use the 1990 amendment benchmarks, because the savings clause in Section 110(n) of the amendments expressly provided for keeping existing SIP obligations in place pending the SIP revision. Thus, the new RFP reduction schedules in the 1990 amendments should not take effect until the EPA approves a new SIP. Until then, they argued, the MTC should be held to the RFP commitments made in the Bay Area Plan but using the updated estimates. In short, plaintiffs continued to insist that the 1990 inventory should be compared to the old RFP lines shown in Figures 4.5 and 4.6.

Plaintiffs also argued that MTC was not following other assumptions and methodologies in the Bay Area Plan. For instance, the agency had failed to correct the figures for the lower ambient temperatures assumed in the Bay Area Plan (53°) and had also assumed higher average road speeds. In addition, plaintiffs complained that the MTC had used incorrect measures for the effectiveness of the I/M program, noting a new report showed that, even with the program enhancements fully in place, the I/M program would still achieve only 18% effectiveness for HCs and just 19% for CO, rather than the 24% figure used in the 1987 inventory.[47] Finally, they objected that the new TCMs would not all be in operation until 1996, whereas the court's order required compliance within 6 months, not 6 years.

Although the MTC agreed to use a higher temperature correction factor in its calculations, it insisted that this would still not cure the apples-and-oranges problem, claiming that "virtually every important assumption" had been updated, and that so many important methodologies had been altered and improved since 1982, when the Bay Area Plan was prepared, that it was "impossible to make fair comparison of emissions levels calculated using one [data] series to emissions levels calculated using the other."[48] MTC went further, though, arguing that it had no obligation to even try to make such a comparison, especially because some of the computer programs used in the original estimates were no longer in use or available:

> Even if it were possible to unravel the updated travel model and emissions assumptions to replicate the old—and it isn't—it is a *non sequitur* to suggest that these advances should be abandoned, and that outdated assumptions and methodologies should be resurrected, in the name of "reasonable further progress."[49]

By this stage in the litigation, the parties' positions with respect to the use of the updated data had solidified. MTC claimed that changes in the complex computer models for predicting traffic and emissions, which had occurred since the Bay Area Plan was developed, made any comparisons to the original RFP standards impossible. Moreover, the agency insisted that if the new estimates were to be used, then they should properly be compared to the revised RFP standards contained in the 1990 amendments rather than the outdated schedules in the original Plan. Plaintiffs, by contrast, maintained that both the 1987 and 1990 inventories measured the same thing. In their opinion, the changes in methodology were merely "evolutionary improvements," which increased the accuracy of the estimates but did not bias the results in any particular way. In fact, the new data represented the "best current estimates" of emissions. The Sierra Club insisted that the MTC's commitments remained enforceable and that it could not ignore the new figures or avoid fully carrying out the Contingency Plan, the "central self-correcting mechanism" in the Bay Area Plan, by claiming that it no longer had access to the original data. The Sierra Club insisted that the MTC be held to account for the much higher current emissions estimates. Plaintiffs pointed out that the fact that the CO design value had been increased to 11.8 ppm, 31% over the 9 ppm standard, was a clear indication that air quality had not improved. They maintained that unless the MTC could show that the present results were overestimated or that the plan targets were underestimated and that the degree of overestimation or underestimation was greater than the RFP shortfalls, then the agency should be held in contempt.

Court Ruling on Contingency Plan

Because no RFP reports had been prepared after 1987, the court had to evaluate the MTC compliance based solely on the material and information provided by the parties. After considering a voluminous amount of highly technical information submitted to the court, and consulting with Dr. Wachs, Judge Henderson concluded that the MTC had shown it could meet RFP for ozone but not for CO. The parties disputed the effectiveness of the original controls and the 16 additional TCMs adopted by the MTC. In some instances, their arguments came down to plus or minus a few tons in emissions reductions, discrepancies that could easily have been exceeded by the statistical errors in the estimates themselves. Nevertheless, the court felt this was the best available data on which to base its decisions and felt obliged to consider it.

Ozone RFP

As we have already noted, in the summary judgment motion, the court had held that the MTC had to achieve both a 430 tpd emissions level for HCs and reductions of at least 85 tpd. Here, the court ruled that this obligation survived the passage of the 1990 amendments so that RFP would continue to be judged by the standards in the Bay Area Plan until the EPA approved a revision. The court also agreed with the plaintiffs that since the 1987 deadline had passed, background emissions had risen and thus the MTC was responsible for even greater emissions reductions. However, the court also accepted the MTC's evidence that by 1991 it could reduce HC emissions by slightly more than 99 tpd. According to the court, the parties had to stay true to the assumptions and methodologies of the original SIP to preserve internal consistency. It therefore adopted the defendants' approach as most consistent with the Bay Area Plan, particularly the Plan's assumptions regarding un-controlled HC emissions levels. As a result, the MTC would not be held responsible for unexpected growth in baseline emissions, but could rely on the 1987 inventory to show compliance.

The court ruled that plaintiffs' reliance on the updated data, to show that HC emissions exceeded the RFP target committed to in the Bay Area Plan, was misplaced, because the 1990 inventory rested on a different set of assumptions and different methodology from the original projections and thus was not comparable to the earlier figures. Specifically regarding HC emissions, the court agreed with the MTC that the 430 tpd limit was relevant only in relation to the Plan's original 1979 baseline data, projecting that HC levels would be 515 tpd in 1987 absent additional controls, and thus con-cluded it would be inappropriate to compare it to the figures generated by the 1990 inventory, which provided a whole new set of projections for use in connection with the SIP revision.

Nevertheless, the court explained that the residual 430 tpd benchmark was not rendered wholly irrelevant because it did take into account *anticipated* growth in emissions occurring after 1987. As described earlier, HC levels would gradually increase following 1987 as population growth began to outstrip any expected improvements in engine technology. Because MTC missed its deadline for compliance, it would have to account for these later increases in any demonstration of RFP, consistent with the Clean Air Act's requirement that SIPs establish a process for achieving and maintaining NAAQS.

Because the court agreed that the Bay Area Plan standards and the 1990 inventories were apples and oranges and could not be compared, the court did not directly address what MTC's responsibility would be if, using a comparable methodology to that used in the Plan, it could be shown that actual emissions were in fact higher than anticipated. The practical effect of the court's holding, however, was that failure to compensate for *unanticipated* growth in pollution would not generate liability under the Clean Air Act but would have to be dealt with under the planning process called for in the 1990 amendments. The EPA has something to say on this matter in its recent guidance regulations for the 1990 amendments and we will discuss their approach in the concluding chapter.

On the merits, the court agreed that, based on anticipated reductions from the existing control measures, MTC could show compliance for ozone by 1991, even without any contributions from the contingency TCMs. The 99.1 tpd in reductions would just cover the 99 tpd shortfall.[50] The court reached a different conclusion, however, with respect to regional CO emissions.

Regional CO RFP

With regard to CO emissions, the court agreed with the plaintiffs that the Bay Area Plan committed the defendants to achieving RFP on the regional and subregional levels, holding that reducing regional CO emissions was a "component of the RFP equation for carbon monoxide."[51] This meant the MTC would have to reduce vehicular emissions to 1,541 tpd throughout the Basin. In contrast to ozone, though, CO emission levels were expected to continue declining over the long term. Thus, instead of having to find additional emissions reductions, the MTC could take advantage of falling pollution levels to reduce the need for further controls. But, even accounting for the decline in uncontrolled emissions between 1987 and 1990 (from 1,934 to 1,892.3 tpd), the court still found that a shortfall existed.

As shown in Table 4.5, which summarizes the court's calculation, by using the 1987 inventory the evidence demonstrated that MTC could at best account only for 336.7 tpd in reductions for the year 1990, leaving it still 14.6 tons short of the RFP target of 1,541 tpd. This was mainly because the evidence did not support the MTC's estimated reductions from the current smog control programs. Using the assumptions and methodologies from the

TABLE 4.5 Court RFP Ruling on CO Emissions

	Emissions
1990 Baseline	1,892.3 tpd
I/M program	−278.9 tpd
Original 10 TCMs	−16.0 tpd
Adopted "2131" TCMs	−41.8 tpd
Subtotal	−336.7 tpd
1991 CO Emissions	1,555.6 tpd
RFP	1,541.0 tpd

Bay Area Plan together with the newly revised lower effectiveness estimate for the I/M program (19%), the court determined that for 1990, reductions equivalent to only 278.9 tpd would be achieved (compared to the original 306.8 tpd estimate used by the MTC).[52] In addition, as noted earlier, several of the "2131" TCMS had apparently become stalled for lack of funding or authorization; therefore, the court credited the MTC with projected reductions of only 41.8 tpd from the remaining measures (down from 74.1 tpd). Adding the minimum expected reductions from the original 10 TCMs (16 tpd) still left a gap. Even with all these control measures, CO emissions still exceeded 1,555 tpd. Thus, the court ruled that RFP for CO for the region had not been met.

On the other hand, the court declined to construe the Bay Area Plan as requiring RFP for CO to be measured by the number of VTs or VMT. The court declared that although future plans could establish enforceable VT and VMT standards as part of a new RFP demonstration, the 1982 Bay Area Plan had not clearly done so.[53]

Areawide and Hot Spot RFP

In contrast to regional CO emissions levels, the MTC did not dispute that the Bay Area Plan measured subregional RFP for CO by reference to the Areawide and Hot Spot benchmarks for San Jose. The court agreed with the MTC that the Bay Area Plan assumed that by 1990 CO levels would have fallen below the 1987 RFP benchmark of 108,000 gm/pk-hr/km^2, and that therefore the record demonstrated that RFP for downtown San Jose had been met. Here again, the court declined the plaintiffs' suggestion to use VTs and VMT as surrogate variables for RFP. The court also held that the Plan established that RFP would be achieved for CO at the Hot Spot at First and Santa Clara by 1990, even without additional controls.[54]

In summary, utilizing the methodology and assumptions of the Bay Area Plan, the court concluded that although MTC had satisfied its obligations with respect to regional ozone and subregional CO emissions, the record did not demonstrate that MTC had made RFP for regional CO emissions, as defined in the Plan. Accordingly, MTC was not in full compliance with the Contingency Plan's requirement to adopt sufficient additional TCMs to "bring the region back within the RFP line." Plaintiffs were therefore awarded partial summary judgment with respect to CO emissions.[55] However, because Judge Henderson acknowledged that "models and projections are by their nature inexact," he declined to impose a precise ton-per-day reduction requirement on the MTC. Instead, he gave the agency 4 months to either identify additional TCMs or explain why it would be infeasible to do so. The court stressed, though, that this meant showing more than mere "inconvenience, unpopularity or moderate burdens."[56]

MTC's Regional CO Compliance Response

In its December 1991 response to the court's ruling, the MTC claimed it could now meet RFP because funding had been restored for several of the stalled TCMs and certain enhancements had been made to other TCMs.[57] The MTC recalculated the effectiveness of the "2131" TCMs, using both the EMFAC 7d program employed in its original estimates for the 1987 inventory[58] and the newer EMFAC 7e program employed in the 1990 inventory and the conformity assessments.[59] Table 4.6 provides a comparison of the revised estimated effectiveness rates for the "2131" TCMs, using both sets of factors. The two programs differ due to changing estimates of the relative contribution of running, cold start, and hot start emissions to the total. These rates, when multiplied by the adjusted baseline emissions left after deducting reductions from the I/M program and the 10 TCMs, produce the estimates of the expected reductions from each measure.

TABLE 4.6 Revised Effectiveness Estimates for the "2131" TCMs

	Percentage of Reductions	
"2131" TCMs	*EMFAC 7d*	*EMFAC 7e*
No. 13	.13	.15
No. 14*	—	.07
No. 15	.43	.38
No. 16	.06	.06
No. 17a	.20	.01
No. 17b	—	.18
No. 18	.05	.04
No. 19	.08	.07
No. 20	.19	.20
No. 21	.03	.04
No. 22	.04	.05
No. 23	.12	.12
No. 24/25*	1.10	.60
No. 26	.48	1.34
No. 27/28	.06	.06
Total	2.97%	3.37%

*Previously stalled measures.

In addition to reductions from the I/M program and the previously adopted contingency measures, the MTC also took credit for the "refunded" TCMs and the enhancements. Although the resulting estimates for individual controls using the two models varied widely, in part because the initial estimates had not projected reductions from some of the TCMs as far into the future as those using the EMFAC 7e factors, overall the estimated emissions

TABLE 4.7 Comparative Emissions Reductions From Alternative Emissions Control Factors, 1991

Control Measure	EMFAC 7d	EMFAC 7e
1991 Baseline	1,878.4 tpd	1,878.4 tpd
I/M program	−278.9 tpd	−278.9 tpd
10 TCMs	−16.0 tpd	−16.0 tpd
Adopted "2131" TCMs	−27.6 tpd	−40.8 tpd
"Refunded" TCMs	−19.3 tpd	−11.4 tpd
TCM enhancements	−16.3 tpd	−10.1 tpd
Subtotal	−358.1 tpd	−357.2 tpd
1992 CO emissions	1,520.3 tpd	1,521.2 tpd

reductions were comparable. The results, which assume an additional 1-year decline in uncontrolled emissions to 1,878.4 tpd, are shown in Table 4.7. In either case, by 1992, residual emissions levels would be below the 1,541 tpd RFP level. Finally, the MTC insisted it had adopted all legally available TCMs and had exhausted its ability to do more because, as a regional planning agency, it had no local land use management authority. Plaintiffs responded, though, that the court had ordered the MTC to identify additional TCMs or else demonstrate the infeasibility of doing so. Instead, they felt the MTC had tried to show RFP had been met by again taking credit for uncertain, potential future reductions to prove present compliance.

Despite the court's earlier ruling, plaintiffs continued to urge that the MTC's analysis didn't cover unanticipated growth, even though MTC admitted that its estimates of VMT were 17% higher than forecasted. The plaintiffs also reiterated their contention that the contingency provisions in the Bay Area Plan were designed to serve as the mechanism to remedy excess travel growth not covered in the original attainment demonstration. They noted that a 1988 report on TCM effectiveness had shown that regional growth was outpacing initial projections, arguing that the Contingency Plan specifically stated that the annual RFP report would "assess the growth in vehicle travel in the region" and thus committed the MTC to adopt additional contingency measures to put the region back on the RFP line.

Court Order Regarding Compliance

The wrangling between the parties finally ended in May 1992, when the court ruled that the CO shortfall had been eliminated through a combination of factors. First, the court agreed that because another year had passed, the MTC could take credit for an additional 13.9 tpd in emissions reductions from the downward sloping CO curve. This reduced total 1991 baseline emissions to just 1,878.4, as shown in Table 4.7, only 337.4 tpd over the 1,541 RFP line (see Figure 4.3).

Second, as also shown in Table 4.7, the MTC was able to demonstrate sufficient reductions from the various programs to achieve RFP by the year 1992. The court adopted the EMFAC 7e emissions factors as being the most recent available information even though its August 19th Order had used the EMFAC 7d figures, reasoning that the plaintiffs did not seriously contest the use of these figures, and the overall result would be the same anyway.[61] Although the court slightly reduced the credit taken for contributions from the I/M program (to 271.2 tpd[60]), the court accepted the additional reductions from the recent funding of three of the previously stalled TCMs (11.4 tpd) and the enhancements made to other TCMs (10.1 tpd). Together with the reductions from the previously adopted "2131" TCMs (41 tpd[62]) and the original 10 TCMs (16 tpd), the combined programs yielded a conservative total of 349.7 tpd in emissions reductions. Subtracting this amount from the reduced baseline left adjusted 1992 net emissions of 1,528.7 tpd, which was still slightly below the target 1,541 tpd figure.[63]

Settlement Agreement

In August 1992, the parties reached a settlement of the lawsuit and the court entered a stipulated judgment. The agreement acknowledged that the MTC had adopted and implemented a satisfactory conformity assessment procedure and had adopted sufficient transportation control measures to bring the region within the RFP line for both CO and HC emissions. Furthermore, the parties agreed that the MTC had properly adopted and applied criteria for reviewing highway projects for delay pursuant to the Contingency Plan. The MTC promised to continue to use the court-approved procedures for assessing the conformity of transportation plans, programs, and projects with the Bay Area Plan until the EPA approved a SIP revision.

Should MTC fail to implement any of the "2131" TCMs, the agency agreed to adopt additional transportation control measures to achieve RFP for regional vehicular CO emissions. Similarly, should those standards not be achieved, MTC also agreed to consider delaying transportation projects in the TIP that have significant adverse impacts on air quality. The court reserved jurisdiction to enforce the parties' continuing obligations.

With the major issues in the case having been resolved in the summary judgment and contempt motions, the parties had strong incentives to settle the lawsuit. CBE and the Sierra Club undoubtedly realized that they had achieved about as much as could be expected from the litigation. They had forced the MTC to adopt some additional control measures and to develop new modeling procedures to assess the impact of its highway construction programs. Had the lawsuit not been brought, it was unlikely that the MTC would have done anything more to get people out of their cars. Given the current state of transportation modeling, the procedures devised and adopted by the MTC and its consultants in response to the lawsuit, and later endorsed by the court, were probably at the edge of technical feasibility. At the time, it was simply not possible to make any more accurate assessments of the potential emissions reductions from the control measures enacted by the MTC and the other defendants. Moreover, political considerations made the possibility of pursuing even more vigorous controls unlikely. The lack of any political entities capable of enforcing measures on a regional basis is clearly an impediment to further progress in this area.

Although the court had basically agreed that the local agencies had done a poor job of carrying out their legal obligations under the Bay Area Plan, on many of the technical issues the court had been flexible and largely deferred to the MTC's administrative expertise. The plaintiffs simply did not have the resources to continue fighting over highly complex modeling issues. Consolidating their victories would certainly have seemed to be a more reasonable course of action.

On the other side, the MTC also chose not to pursue the matter any further. Like the plaintiffs, the agency was probably exhausted from the legal fight. In addition, given the court's willingness to delay highway projects until adequate procedures were in place, further litigation might only jeopardize its construction program. Although the agency could have appealed the district court's ruling, its appeal of the court's injunction had already been rejected by the Ninth Circuit, and the prospect of reversing Judge Henderson's decisions on the other matters was not strong. The MTC had also expended a substantial amount of money hiring outside counsel to defend their practices, a sum that would only grow if the case was continued. In any event, the agency seemed to have accepted the fact that its existing procedures were inadequate and that it would have to pay closer attention to its obligations under the Clean Air Act in the future. Until a new SIP was approved, the MTC acknowledged it would have to follow a two-part track—adhering to the commitments in the Bay Area Plan while at the same time preparing a new cleanup plan that would meet the requirements of the 1990 amendments. Inasmuch as it would soon be replacing the original plan anyway, the matter could hardly have seemed worth fighting over. The agency also appeared to be comfortable with the new modeling procedures it had developed, and its ability to demonstrate attainment within the time frames of the new legislation.

Although the experience no doubt left both sides somewhat bitter from what each party perceived as the intransigence of the other, it also gave them a much better understanding of the possibilities and limitations of travel modeling. In this sense, the case was a valuable lesson to all involved in how far the state of transportation planning has come and how much yet needs to be done.

Notes

1. *Citizens for a Better Environment v. Deukmejian* (CBE I), 731 F. Supp. 1448, 1461-1462, 31 Environment Review Cases (ERC) 1213, 1225 (N.D. Cal. 1990).

2. Draft Implementation Plan for the Contingency Plan in the 1982 Plan (July 28, 1989, revised October 20, 1989), at 4 [hereinafter Draft Implementation Plan].

3. EPA letter to the MTC, dated September 1, 1989.

4. Draft Implementation Plan, supra note 2, at 3.

5. New facilities would include new highways, lane additions, and upgrading facilities. Examples of capacity-enhancing projects include restriping projects to add a through lane, truck climbing lanes, passing lanes, and auxiliary lanes. Rehabilitation projects consist of projects such as roadway and bridge reconstruction, maintenance and equipment facilities, and protective barricades. Operational improvements having no significant air quality impacts would include safety improvements, noise abatement, rest areas, and highway planting. Metropolitan Transportation Commission Resolution No. 2107 (October 30, 1989, revised December 20), 1989, at 2-4.

6. The MTC required the following information relative to the development of the HC estimates:

- Source of the demographic and traffic projections (ABAG/MTC or other) and the year projections were prepared;

- Traffic volumes and speeds for project alternatives for the future planning year; also traffic volumes and speeds on adjacent routes where significant diversion of traffic is expected to occur; estimates of vehicle delay should be provided if crucial for choosing among project alternatives;
- The environmental document should contain an explanation of methodology used to calculate HC impacts; the impacts should be calculated by considering not only the performance of the facility that composes the project but also that of adjoining and parallel facilities whose performance may be either positively or negatively affected by the project.

Id. at Appendix 1 to Attachment B, ¶ 2a.

7. The MTC required the following information regarding the development of the CO estimates:

- Source of the demographic and traffic projections (ABAG/MTC or other) and the year projections were prepared;
- Traffic volumes and speeds for project alternatives for the future planning year; and estimates of vehicle delay should be provided if crucial for choosing among project alternatives;
- CO background levels and any "rollback" assumptions for future planning years;
- Source for composite automobile emission rates (such as the EMFAC7D emissions table produced by the Air Resources Board (ARB);
- Analytical approach used such as the CALINE3 and CALINE4 dispersion models developed by the California Department of Transportation.

Id. at ¶ 1a.

8. Exempt activities were covered by Environmental Protection Agency/Federal Highway Administration (EPA/FHWA) Region IX Screening Criteria for projects affected by Section 176(a). These included safety programs, mass transit, and projects to improve air quality. Draft Implementation Plan, supra note 2, Attachment A. Section 176(a) was repealed by the 1990 amendments.

9. EPA letter to the MTC, dated September 1, 1989.

10. Candidate TCMs were broken down into a number of categories, which included (a) marketplace strategies (gas tax and bridge toll increases, differential parking pricing, auto free/no parking zones, employer travel allowances, and increased subsidies for transit and carpools/vanpools), (b) enriching commuter options (transit expansion, increased ride sharing and trip planning/coordination, bicycle paths and storage facilities, and pedestrian-oriented urban design policies), (c) congestion management (ramp metering, signal timing, truck delivery time regulations, and driver education), (d) regulating driving (assigned driving days, phasing out older cars, limiting car ownership/registration and limiting parking), (e) auto facilities Rrstrictions (limiting further highway expansion and parking facilities), (f) technological fixes (engine redesign, alternative fuels, telecommuting), (g) public education and information, and (h) growth management and land use ("balance growth" ordinances, jobs/housing balance, improved pedestrian access/site design, mixed use developments, and tax revenue sharing). Draft Implementation Plan, supra note 2, Table 3.

11. EPA letter to the MTC, dated February 14, 1990 (emphasis added).

12. CBE I, 731 F. Supp. at 1460, 31 ERC at 1223.

13. Rejected TCMs included increasing commuter parking surcharges, higher Bay Bridge tolls, and HOV lane designations.

14. In the alternative, the Sierra Club asked for summary judgment that the MTC was not in compliance with the contingency requirements of the Bay Area Plan.

15. EPA regulations for extending the deadline to 1987 called for straight-line reductions in emissions beginning in the year 1979 to the year of attainment. Final Policy—Criteria for Approval of 1982 Ozone and Carbon Monoxide Plan Revisions for Areas Needing an Attainment Date Extension, 46 Fed. Reg. 7182, 7187 (January 22, 1981); see also General Preamble for Proposed Rulemaking on Approval of Plan Revisions for Nonattainment Areas, 44 Fed. Reg. 20,372, 20,377 (April 4, 1979). The 1982 Bay Area Plan RFP schedule for ozone met these requirements.

16. The parties disputed whether the MTC was responsible only for shortfalls in the transportation sector or had to make RFP overall. The Bay Area Plan states,

> In July of each year, an RFP report will be submitted to the EPA. Part of this report will be a review of the status of implementation of the adopted TCMs. A second portion would assess the growth in vehicle travel in the region. If a determination is made that RFP is not being met for the transportation sector, the MTC will adopt additional TCMs within 6 months of the determination. These TCMs will be designed to bring the region back within the RFP line.
>
> Bay Area Air Quality Plan (BAAQP), at 107.

17. Revised estimates of the I/M program suggested that it was only 16% effective for 1988 but projected it would be 28.3% effective in the 1990s. California I/M Review Committee, *Evaluation of California Smog Check Program, Second Report to the Legislature,* May 1989, p. 7, Fig. 1. Interpolating these figures suggested that the program would be 18% effective in 1989. Estimated reductions for the improved I/M program were calculated by applying the revised percentage effectiveness figures to the (increasing) estimated annual baseline HC vehicle emissions inventory subject to inspection as follows:

Year	1987	1988	1989	1990	1991	1992	1993
HC Inventory	116	120	122	125	127	129	132
% Effective	14.1	16.0	18.0	20.0	24.0	28.3	28.3
Reductions	16.4	19.2	22.0	25.0	30.5	36.5	37.4

Declaration of Edward Miller, June 20, 1990, Exhibit G.

18. The MTC estimated that by 1989 a total of 44.9 tpd in reductions could be achieved from (a) stationary source controls in the Bay Area Plan (35.7 tpd) and (b) previously adopted contingency measures (9.2 tpd). These, combined with 3.7 tpd in reductions from the four court-mandated controls, produced a total of 48.6 tpd in reductions. See Table 4.3 in the text.

19. *Citizens for a Better Environment v. Deukmejian* (CBE II), 746 F. Supp. 976, 981, 32 Environment Review Cases (BNA) 1136, 1140 (N.D. Cal. 1990). The court stated that the commitment was a joint obligation of the MTC, the District, and ARB and that the MTC had already been ordered to implement the transportation portion of the Contingency Plan. Id. at 984, n.15, 32 ERC at 1142, n.15.

20. *Citizens for a Better Environment v. Wilson* (CBE III), 775 F. Supp. 1291, 34 ERC 1369 (N.D. Cal. 1991).

21. Id. at 1299, 34 ERC at 1376.

22. Id. at 1294, 34 ERC at 1371. Following the 1990 amendments, Section 171(1) now reads,

The term "reasonable further progress" means such annual incremental reductions in emissions of the relevant air pollutant as are required by this part or may reasonably be required by the Administrator for the purpose of ensuring attainment of the applicable national ambient air quality standard by the applicable date.

42 U.S.C.A. § 7501(1) (West 1995).

23. 42 U.S.C.A. § 7410(n)(1) (West 1995). The court also rejected the MTC's rather strained argument that the RFP commitments in the Bay Area Plan somehow ceased to exist after the 1987 compliance deadline, leaving the Bay Area without any enforceable air pollution standards to be affected by the savings clause in Section 110(n). Indeed, the EPA had specifically directed the MTC to adopt sufficient TCMs to eliminate the 1987 CO and HC shortfalls. CBE III, 775 F. Supp. at 1297, 34 ERC at 1374.

24. The MTC's countermotion for summary judgment was therefore denied and plaintiffs' motion for partial summary judgment on this issue was granted. CBE III, 775 F. Supp at 1299, 34 ERC at 1376.

25. The MTC also argued that it met its share of reductions from any increased growth in baseline emissions from 1987 to 1989. The agency asserted that it was required only to make RFP for the transportation sector and was not obligated to make it for all sectors even though the District, which had authority over stationary sources, had not adopted any contingency measures. See note 16 supra. Because the transportation sector was accountable for only 29 tpd of the originally required 85 tpd reduction, or 34%, the MTC argued that at most it would be responsible only for covering 34% of the 99.1 tpd required due to growth—or 33.69 tpd. Because it projected reductions of 30.5 tpd from the I/M program, this left only 3.19 tpd unaccounted for, which would be less than the projected 3.83 tpd from the "2131" TCMs.

26. The I/M program did not cover heavy-duty trucks, buses, or motorcycles, which contributed the remainder.

27. Association of Bay Area Governments, *Bay Area Air Quality 1987 RFP Report*, November 1988, at 40. HC emissions were reduced only 0.6 to 2.0 tpd instead of the 2.69 tpd predicted in the Bay Area Plan, and only 1.5 to 1.9 tpd for NOx instead of 2.79 tpd. Id. at 41-42. See Table 2.1 in the text.

28. Association of Bay Area Governments, *Bay Area Air Quality RFP Reports: A Retrospective Assessment of 1985 and 1986*, July 1988 [hereinafter 1988 Retrospective]. According to this report, in 1986 the I/M program was only about 10% effective for CO emissions (1,468 × .098 = 144 tpd), instead of the anticipated 25%. The report predicted program effectiveness would improve to 16.3% in the short term and to 25.2% over the longer term. Reductions of only 107 tpd had been achieved in 1985, according to the same report.

29. Metropolitan Transportation Commission, *Bay Area Travel Forecasts: Year 1987 Trips by Mode, Vehicle, Miles of Travel and Vehicle Emissions, Technical Summary*, August 1990 [hereinafter 1987 Travel Forecast], Tables C-4 and C-5.

30. The MTC initially assumed a 21.6% effectiveness rate and calculated the reduction amount as follows: 1,468 tpd × .0216 = 316.38. This figure, however, was arrived at using a flawed methodology. First, the MTC assumed that the projected emissions reductions for CO would be the same as for HCs (25%). Second, it applied the wrong percentage reduction factor, based on an effectiveness rate of 86% derived from comparing observed HC levels for 1990 (25 tpd) to the RFP target (29 tpd). See note 17 supra. Thus, 25% × (25/29) = 21.6%. Revised estimates showed the I/M program was only 18% effective for CO in 1988 and would increase to just 26.9% in the 1990s. California I/M Review Committee, *Evaluation of California Smog Check Program, Second Report to the Legislature*, May 1989, p. 7, Fig. 1. Interpolating these figures indicated a 20.9% effectiveness rate in 1990, resulting in a figure of 306.8 tpd. In its supplemental submissions to the court, MTC corrected its earlier approach and recalculated the residual emissions as follows:

Year	1989	1990	1991	1992	1993
CO Emissions	1,468	1,468	1,468	1,468	1,468
% Effective	19.4	20.9	23.8	26.8	26.9
Reductions	284.8	306.8	349.4	394.9	394.9

Supplemental Brief by the MTC in Opposition to Motion for Contempt and Summary Judgment, July 22, 1991, at 22, n.12.

As explained in notes 52 and 60 infra, however, the figures were still too high. First, the effectiveness estimates turned out to be too optimistic. Second, MTC applied the reduction factor to a constant base (1,468 tpd) even though CO emission levels were declining over time. The court eventually corrected these errors.

31. BAAQP, at 148-150, Figures 22 and 23.

32. 1987 Travel Forecast, supra note 29, at 14, and Tables 1 and 7. Plaintiffs estimated the "surrogate" RFP number of noncommercial VTs at 11,110,730. Projected VTs in 1987, based on the MTCFCAST model and off-model estimates for home-to-school, interregional, and truck travel, were 14,283,821 trips, of which 1,692,900 trips were due to truck traffic.

33. The EPA concluded that the Contingency Plan would have to account for shortfalls caused by higher than anticipated growth in VMT, VTs, and congestion levels in the Bay Area. On the other hand, the EPA agreed that the MTC could take credit for additional reductions from the enhanced I/M program to correct RFP shortfalls, and that the TCMs need not bring the areas into ozone and CO attainment but had to at least reduce emissions in the transportation sector to the levels projected in the Bay Area Plan. Letter from the EPA to the MTC, February 14, 1990, Attachment A, pp. 1-2.

34. 42 U.S.C.A. § 7512a(b)(2) (West 1995).

35. 42 U.S.C.A. § 7512a(a)(2)(A) (West 1995).

36. The projected 9.1% reduction in Areawide CO consisted of 5,760 g/pk-hr from the I/M program (4.9%) and an additional 5,040 g/pk-hr from the Commute Transport Program (4.2%). 1988 Retrospective, supra note 28. See BAAQP, at 138, Table 31.

37. 1988 Retrospective, supra note 28, Table 1. In 1986, Areawide reductions from the I/M Program were only about 4,200 g/pk-hr/km^2 and from the Commute Transport Program only about 4,000 g/pk-hr/km^2. Hot Spot reductions from the I/M program were only 1,500 g/pk-hr.

38. Extrapolating from the trend line, the uncontrolled CO emissions would be 105,300 g/pk-hr/km^2 for 1990. MTC claimed reductions from the I/M program operating at 86% of design efficiency for a total of 4,965.5 g/pk-hr/km^2, plus 526.5 g/pk-hr/km^2 for the signal synchronization program, and 4,212 g/pk-hr/km^2 for the Commute Transport program for a total reduction of 9,704 g/pk-hr/km^2, leaving 95,596 g/pk-hr/km^2. Metropolitan Transportation Commission's Memorandum of Points and Authorities in Opposition to Motion for Contempt or Summary Judgment, March 1, 1991, at 29-30 [hereinafter MTC's Opposition Memorandum].

39. Id. 31, n.7. the MTC calculated the I/M program would reduce local CO emissions an additional 1,759.5 grams for a total of 4,168.6 g/pk-hr/km^2. As with regional and Areawide emissions, The MTC assumed that the I/M program would be operating at 86% of design effectiveness. Because the Bay Area Plan called for the smog check effort to reduce Hot Spot emissions from 8,474 g/pk-hr to 6,428 g/pk or 2046 grams at full efficiency, the MTC obtained its estimate as follows: 2,046 g/pk-hr × .86 = 1,759.5 g/pk-hr. Emissions reductions from the Commute Transportation Program were not calculated for the Hot Spot location.

40. Id. at 14, n.6. See CBE II, 746 F. Supp. at 980, n.7 and 983.

41. Supplemental Brief by Metropolitan Transportation Commission in Opposition to Motion for Contempt and Summary Judgment, July 22, 1991, at 23, n.13. In it motion for partial summary judgment, the MTC argued that it should be allowed to use this approach instead of following the RFP analysis based on the 1977 Clean Air Act.

42. 42 U.S.C.A. § 7512(a)(1) (West 1995).

43. The new design value was based on the highest of all second-highest measurements of CO emissions taken at all of the monitoring stations throughout the Bay Area in 1988 and 1989. The specific measurement was recorded in November 1989. The MTC initially calculated the reductions from a 1990 baseline (1,698.3 tpd). Plaintiffs objected to the MTC moving the CO baseline back one year, to 1989, to give itself more time to reach the new 23.7% reduction standard. Using the 1990 base figure produced only a 24% reduction by the target date, which would be just above the new RFP standard. By stretching out the compliance period, MTC was able to claim even higher percentage reductions.

44. These were No. 14 ($2 Bay Bridge toll), No. 23 ("employer audits" to help employers develop commute programs for employees), and Nos. 24/25 (expanding and maintaining signal timing programs). See Table 4.2 in the text.

45. A later estimate showed the figures to be 160,260 g/pk hr/km^2 for 1989 and 122,286 g/pk-hr/km^2 for 1995, a 23.7% reduction. The MTC also claimed that it should be allowed credit for the Oxygenated Fuels Program that would be required under the 1990 amendments. With the Oxygenated Fuels Program in place, the estimate dropped to 111,700 g/pk-hr/km^2. Second Supplemental Declaration of Greig W. Harvey in Opposition to Motion for Contempt or Summary Judgment, April 26, 1991.

46. As for any failure to adopt other possible TCMs, the MTC insisted that the Section 108(f) list applied only to SIP approvals and not to contingency measures. See *Delaney v. EPA*, 898 F.2d 687, 692 (8th Cir. 1980). Moreover, it reiterated that the battle over unadopted TCMs was academic because RFP could be established no matter how it was calculated.

47. California Air Resources Board, *Development of the CALIMFAC California I/M Benefits Model*, January 1991.

48. Metropolitan Transportation Commission's Reply Memorandum in Response to Order of June 25, 1991, July 25, 1991, at 2. MTC claimed that it was not clear what temperature was assumed in the Bay Area Plan and argued that in any case it did not need to follow the same painstaking procedures, used in the Bay Area Plan, for an attainment demonstration to show RFP. The agency believed that the appropriate temperature to use would be the one on the day that the new design value was recorded, 67°, but the MTC agreed to split the difference and use 60°. Supplemental Brief by Metropolitan Transportation Commission in Opposition to Motion for Contempt and Summary Judgment, July 22, 1991.

49. Metropolitan Transportation Commission's Reply Memorandum in Response to Order of June 25, 1991, July 25, 1991, at 5.

50. Although the projected 30.5 tpd reduction in the transportation sector by 1991 was probably incorrect, due to revised estimates of the effectiveness of the I/M program from 23.8% down to only 18%, nevertheless, the court found insufficient evidence that RFP for ozone had not been met, noting that improving reductions in the stationary source category would also contribute to attainment. CBE III, 775 F. Supp. at 1302-3, 34 ERC at 1379.

51. Id. at 1303, 34 ERC at 1380. This was also the EPA's position. In a letter directing the MTC to correct its RFP shortfalls, the director of the EPA wrote, "These include *not only regional HC and CO levels,* but also Hot Spot and Areawide CO emissions for the San Jose area." EPA letter to the MTC, dated February 14, 1990 (emphasis added). See 775 F. Supp. at 1304, 34 ERC at 1380.

52. Based on the revised estimate of 19% effectiveness, the court recalculated the figure as follows: 1,468 tpd × 0.19 = 278.9 tpd. Although the 1991 CALIMFAC Report projections (see note 47 supra) were based on a model, the court found it to be the most recent reliable information

on the effectiveness of the I/M program, particularly in light of the fact that actual data would not be available until 1995.

53. The EPA has rejected a similar interpretation of Section 182(d)(1)(A) to require certain areas to offset any growth in VMT above 1990 levels, rather than just offsetting VMT growth only when it leads to actual emission increases, stating,

> Because VMT is growing at rates as high as 4 percent per year in some cities such as Los Angeles, these cities would have to impose draconian TCM's such as mandatory no-drive restrictions to fully offset the effects of increasing VMT if the areas were forced to ignore the beneficial impacts of all vehicle tailpipe and alternative fuel controls.
>
> General Preamble for the Implementation of Title I of the Clean Air Act Amendments of 1990, 57 Fed. Reg. 13,498, 13,522 (April 16, 1992).

54. See 1982 BAAQP, p. 138, Table 31.

55. The court refused to find the MTC in contempt of its September 19, 1989, order because the order was not sufficiently definite with respect to MTC's obligation to adopt sufficient additional TCMs to warrant such a finding. CBE III, 775 F. Supp. at 1300, 34 ERC at 1377.

56. Id. at 1307-1308, 34 ERC at 1384.

57. These were No. 14 (bridge toll increases), No. 15 (gas tax hike), and Nos. 17a and 17b (expanded ferry and BART service) and funding for the Guadalupe Light Rail Transit. See Table 4.2 in the text.

58. The original estimates were in error because the MTC had applied the EMFAC 7d effectiveness percentages to a base of 2,400 tpd, which was considerably higher than the adjusted base assumed under the Bay Area Plan.

59. The most appropriate model to use would have been the EMFAC 6c, which was used to develop the Bay Area Plan, but according to the MTC, the computer program for this model was no longer available.

60. Order, Case No. C89-2044 TEH, May 11, 1992, at 5, n.4. The court used this figure instead of 278.9 tpd to reflect the fact that the I/M program would be operating on a smaller CO baseline inventory in the year 1991 (1,427 tpd) compared to 1987 (1,468 tpd). Thus, 1,427 × 0.19 = 271.2.

61. Id. at 6, n.5. Although plaintiffs complained that by using the 1997 effectiveness estimates the MTC was taking credit for improvements from TCMs that might not be realized for several years, the court considered it reasonable to do so because there was no specific timeline in the Contingency Plan for demonstrating compliance, and in any event, the schedule showed that RFP would be attained by 1994.

62. The percentage rate for the adopted "2131" TCMs (when fully implemented as of 1997) was 2.58%. This figure is applied to the now slightly higher adjusted baseline reflecting the smaller reductions from the I/M program: $1,878.4 - (271.2 + 16) = 1,591.2$. The court therefore recalculated the reductions from these TCMs as follows: $1,591.2$ tpd × $.0258 = 41.0$ tpd. Id. at 6, n.5.

63. The MTC also claimed credits for its new Oxygenated Fuels Program. The 1990 amendments required all CO nonattainment areas with design values above 9.5 ppm to adopt an oxygenated fuels program requiring 2.7% oxygen by weight in gasoline. In December 1991, California adopted a program calling for a 1.8% to 2.2% reduction, effective November 1992. The MTC intended to apply for a waiver of the federal standards on the grounds that the higher oxygen content would interfere with achieving ozone and fine particulate matter (PM-10) standards. In light of its holding that RFP was demonstrated anyway, the court declined to reach the issue of the MTC's authority to adopt additional TCMs or to take credit for the Oxygenated Fuels Program. Id. at 11, n.9.

Analysis and Conclusions

The district court made a number of important rulings in this case regarding both conformity determinations and reasonable further progress (RFP). First, it held that public interest plaintiffs could sue to enforce the State Implementation Plan (SIP), including its contingency and conformity provisions. The defendant agencies could not be challenged, though, simply over their failure to achieve the federal air quality standards. The Clean Air Act was designed so that local agency planners, with public input, would determine the best approaches to meeting the federal clean air goals. Once the plan is finalized and approved, though, Congress did not want outside parties relitigating either the propriety of the methods chosen or whether those methods would in fact achieve the National Ambient Air Quality Standards (NAAQS). It did, however, intend for agencies to be held to the commitments made in their SIP. The court's decision to hold regional agencies responsible for the commitments in their plans is certainly the most significant aspect of the case, though it may also have the unintended side effect of causing planners to be more circumspect in the commitments they make in those plans.

Second, Judge Henderson found that the contingency provisions in the Bay Area Plan obligated the Metropolitan Transportation Commission

(MTC) and the other defendants to take specific steps in the event that reasonable progress was not made toward meeting air quality goals. In the case of mobile sources of pollution, the court ruled that the Plan required the defendants to adopt additional control measures to bring the region "back within the RFP line." The court interpreted RFP to incorporate any anticipated increases in regional emissions after the 1987 federal deadline but did not require the MTC to compensate for any unexpected increases in ambient pollution levels due to growth. In other words, RFP could be demonstrated by subtracting the anticipated emissions reductions due to the various control measures from the initial baseline assumptions in the 1982 Plan. The court eventually accepted the MTC's analysis of the effectiveness of its additional transportation controls as likely to provide sufficient emissions reductions to satisfy the RFP standards.

Third, the court held that the MTC's *qualitative* conformity assessment procedures did not satisfy the agency's commitments in the Plan or fulfill the requirements of the Clean Air Act. The court concluded the defendants had to quantify emissions levels from all proposed transportation plans and programs to determine whether they conformed to the approved SIP. The decision affirmed the principle that valid conformity determinations may not be made, and projects may not proceed, where doing so would lead to a violation of the clean air standards or otherwise prevent the region from maintaining RFP. For the MTC's 1997 model, those assessments would have to be consistent with the revised attainment standards in the 1990 amendments. Because neither the amendments nor the 1982 Bay Area Plan specified how to determine conformity with respect to the transportation plan's long-term horizon year, however, the court concluded that the MTC's proposal for a "build"/"no-build" comparison with respect to its year 2010 model was not unreasonable. The court also ruled that the MTC's computer model properly took into account the effects of highway expansion and congestion reduction strategies on the distribution of growth and development in the region, and that it did not have to consider the potential impact of transportation investments on the level of future growth in the region.

Finally, the court ruled that the MTC was not obligated to perform separate conformity assessments on individual highway projects, but would at least have to undertake some preliminary qualitative assessment to identify potential CO hot spots in the transportation improvement program (TIP), although individual project hot spot assessments could wait until the environmental review stage, when the projects would be more clearly defined.

The fact that the court agreed to enjoin specific highway construction projects until acceptable conformity assessment procedures were in place is also noteworthy. Considering the fact that the 1990 amendments have tightened the requirements for monitoring air quality impacts from highway projects, the willingness of the federal courts to approve an injunction remedy is particularly significant.

Regardless of whether this lawsuit had its intended impact, efforts to improve air quality in the San Francisco Bay Area have been paying off. As of 1994, the region met the national criteria pollutant standards for ozone and carbon monoxide (CO). On April 24, 1995, the Environmental Protection Agency (EPA) officially redesignated the Bay Area as an attainment area for ozone, the largest metropolitan area in the nation to date to meet the federal standard.[1] The change came after the area logged 3 consecutive years with an average of no more than 1 day over the 12 parts per hundred million (pphm) 1-hour standard and adopted a long-term maintenance plan.[2] The region has also attained the federal 8-hour CO standard and has petitioned the EPA to be designated an attainment area for this pollutant. All this came despite the fact that the number of motor vehicles has doubled from 2 million to 4 million since 1965 and population has increased almost 50%, from roughly 4 million to 6 million. Vehicle miles traveled in the Bay Area have also increased from about 48 million miles to 123 million in the same period.[3]

Clearly, the questions posed in this lawsuit, particularly over the impact of highway construction on regional growth, may also affect transportation planning in other regions trying to comply with the provisions of the 1990 Clean Air Act Amendments. It is too early to say how courts in other circuits will follow this court's ruling; but given the new emphasis in the 1990 amendments on controlling growth-induced pollution, discussed below, transportation planners would be well-advised to begin adding the sorts of assessment tools developed in the Bay Area to their own models.

The Meaning of the Court Ruling for Transportation Planning and Modeling

The court rulings in the Bay Area litigation generally had profound implications for transportation planners, who realized that it marked a fundamental transition in the ways in which transportation planning methods are used in practice. The MTC had used standard transportation planning methods that are in general use throughout the world; in fact, MTC was nationally acknowledged to be among the most up-to-date—even innovative—agencies of its type. If its planning procedures were found to fall short of the requirements of the Clean Air Act Amendments, then it follows that most planning agencies were vulnerable to similar challenges. Throughout the preceding 30 years or more, analytical techniques in transportation planning were used in a particular way and planners were comfortable with that particular role for analytical models.[4] The Bay Area case forced planners to recognize that their methods might now have to be viewed in a different light. Indeed, the case made many planners question whether existing models have the capabilities to be used with reasonable precision in the manner that the Clean Air Act Amendments imply that they should be used.

Since the resolution of the case, planners have been reconsidering the role of forecasting models in the transportation planning process and interpreting the implications of the case for their body of professional expertise. The Transportation Research Board, an arm of the National Research Council, appointed an independent panel of 16 experts in the fields of transportation planning, environmental science, and air quality policy to examine the basic question of whether expanding metropolitan highways generally contributes to improved air quality by eliminating bottlenecks and congestion, or generally contributes to worsened air quality by inducing increases in travel. The final report of this study acknowledges that it was motivated in part by the outcome of the Bay Area lawsuit.[5] In addition, the federal government has embarked on a major new research program aimed at producing the next generation of planning models, also citing the Bay Area case as justification for that effort. The next section explains some of the limitations of current travel demand models. Following that section, we begin to examine the ways in which the court's interpretations of the Clean Air Act Amendments place new demands on planners to produce model results that overcome the limitations of the traditional methods.

**Limits to the Application of
Transportation Planning Models**

As discussed in the Chapter 1, travel demand forecasting models were originally developed to support large-scale transportation planning and construction in the late 1960s and early 1970s. Planners then were more concerned with the size and patterning of high-capacity freeway systems necessary to accommodate the federal government's postwar program of suburbanization. The standard model that emerged was actually based on a series of separate computer models. Each step in the sequence of models required particular data and certain assumptions about human behavior, and the errors resulting from inadequacies in one stage of the analysis were carried forward into the later stages.[6]

Research and development to expand the capabilities of transportation modeling during the 1970s and 1980s tended to add sophistication to each step in the sequential independent modeling process. Later models accommodated larger numbers of geographic zones, more variables, and more complex mathematical functions, but the basic structure involving a sequence of separate models remains intact in most of the models widely used in practice today, including the models employed by the MTC in the San Francisco Bay Area. For example, by the late 1970s and early 1980s, many regions began to consider major capital investments in rail transit systems to complement their highway networks or even as substitutes for previously planned highway facilities. In response to the need to plan regional rail systems, planners devised far more effective and sensitive modal choice models to estimate transit ridership under a variety of different conditions, although the modal choice model typically remained only one in the sequence of independent models.

An understanding of the important limitations of existing travel forecasting modeling technology emerges from this history of travel demand models. The question of feedback among the models, which became an issue in the Bay Area lawsuit, is one good example. If each model in the sequence is really independent of all the others, the models lack realism in part because of the absence of feedback, which in practical terms means that the earlier models in the sequence are not influenced or affected by the later models in the sequence. For example, it is logical to assume that travelers might, under conditions of extreme traffic congestion on one route, choose another; or they might choose to use transit rather than to drive if a critical artery is congested; or they might decide to change the time of their trip to a different hour of the day, in the hope of encountering less congested conditions; or they might decide to visit a different destination than they would if less congestion existed. The modeling process, however, does not provide for such adjustments unless feedback exists from later models in the sequence to earlier ones. If the route choice model follows the destination choice model in the sequence, and if both follow the mode choice model, the models may not allow either route choice or mode choice to be influenced by congestion levels within the modeling process.

Furthermore, if land use patterns do indeed adjust to new transportation capacity, similar feedback is needed to analyze the dynamic equilibrium that will evolve over time as both travel and land use influence one another. Over the years, modelers have included an increasing number of interconnections and feedback loops between the models to more closely approximate the kinds of adjustments in travel that would seem to occur in real life. But those adjustments to the model set, which are intended to make the results more realistic, also make the models more complex, lead to increased data requirements to make them work, and make it difficult to estimate the ways in which modeling errors could accumulate through different applications of the models.

Other adjustments to the set of models are needed to evaluate the effectiveness of transportation demand management strategies (such as the pricing of parking or increases in bridge tolls), which became more important to policy debates as transportation control measures were included in regional air quality management plans and as the analysis of their potential impacts was needed as part of conformity assessments. The original travel demand models incorporated relatively few price variables, because they were intended to support a policy of addressing traffic congestion by expanding highway capacity rather than by depressing travel demand through price increases.

It is also important to acknowledge that the validity of any mathematical model is greatest when evaluating alternative sets of conditions for which data exist and under which the model was originally developed. In the jargon of planners: Models are valid only over the range of conditions for which they were calibrated. In practical terms, however, the use of these models in policy making often pushes planners to applications that go far beyond these conditions. For example, if a mode choice model is developed in a region in which 3% to 5% of all trips are made by transit, the forecasts made by these models under conditions intended to lead to substantial increases in transit use—to, say, 15% or 20% shares of the market—must be suspect. If the models have never been tested in situations in which 15% or 20% of all trips are made by transit, use of the models to make such forecasts is subject to errors of unknown magnitudes. Similarly, if the responses of travelers to bridge tolls have been modeled on the basis of tolls ranging from 25 cents to $1, forecasts of the effects of bridge tolls in the range of $3 to $5 must be regarded as highly speculative and also prone to errors in estimation.

The Accuracy of Forecasts Made
by Transportation Planning Models

Given these many concerns about the limitations of the planning methods and their particular evolutionary path, it is reasonable to ask how accurate those models can be, and in fact this question was of great importance to the court. Although the question may be a reasonable one, it is one that cannot be easily answered. In part, it is difficult to estimate the accuracy of planning models because the accuracy of a model can be judged or measured in many different ways

First, it is important to distinguish between *precision* and *accuracy*. The distinction may be illustrated simply, by way of an example. If a process of modeling mobile air pollutant emissions estimates that one potential transportation control measure will reduce pollutant emissions by 3.5 tons, it would seem that the level of *precision* of the model is approximately 0.1 ton. That is, the model would seem to be able to predict emissions reductions that differ under different circumstances by as much or as little as 0.1 ton. The model that is capable of estimating a reduction of 3.5 tons might also produce an estimate under different circumstances of 3.4 tons or 3.6 tons; but it is not reasonable to expect another test using the model to produce an estimate of 3.500001 tons, a far more precise estimate. The precision of the estimate, however, says little about its *accuracy*. For example, the model could produce an estimate of 3.5 tons with a standard error of 0.1 ton, meaning that about two thirds of the time we would reasonably expect the "true" value of the estimated quantity to lie within 0.1 ton of the estimated value; there would be two chances in three that the true value of the predicted quantity would lie between 3.4 and 3.6 tons. The modeling process could also produce an estimate of 3.5 tons with a standard error of 1.0 ton, meaning that there are two chances in three that the true value would be between 2.5 and 4.5 tons. It could also produce an estimate of 3.5 tons with a standard error of 10 tons, meaning that there are two chances in three that the true value would be between –6.5 tons and 13.5 tons. The estimate having the standard error of 0.1 is far more accurate than the estimate having the standard error of 10 tons. This is a serious issue in transportation and air quality modeling. It is quite possible to produce results that have great precision but have far less accuracy.

Transportation and air quality models operate simultaneously on many geographical zones and incorporate many roadway segments. A large-scale travel demand model for an area such as the San Francisco Bay Area encompasses hundreds of geographic zones, and tens of thousands of highway segments. How shall we judge the accuracy of such a model? One way would be to compare the flows of traffic predicted by a model on certain roadway segments with actual flows as revealed by traffic counts. Similarly, we could measure pollutant levels at particular locations and compare those with levels predicted by the models. If we were to do this, however, we would again face many questions in defining the accuracy of the models. Using traffic flows as a criterion measure for accuracy, for example, we might treat all road segments equally, but we might note that some roadway segments carry 10 or 100 times the volumes of others, and it might turn out that the models forecast more accurately for high-volume roads than for low-volume roads, or vice versa. It might, similarly, turn out that because of some systematic error in the models, its forecasts are systematically more accurate in one sector of the city than another—downtown, for example, versus the suburbs. It also could easily happen that although, on average across all highway segments, the model produced a 10% error between forecast and observed traffic flows, it could achieve this by producing less than a 1% error on some segments and more than a 100% error on others. If the transportation control measure under study relies heavily on a segment of the highway network for which the model predicts poorly—a particular bridge or tunnel, for example—the analyst will rightfully place far less confidence in the model's validity than its average error might suggest is warranted.

There are additional questions about the accuracy of travel forecasting models that are relevant to the matters considered by the court. Because the models are used to forecast traffic and air pollution some 5 or 10 or 20 years into the future, the accuracy of the models in their forecasting role can logically be tested only by waiting for the passage of that much time to see how well they forecast. Policy, however, must be made today, and planners do not have the luxury of assessing the accuracy of their models under the conditions for which they were developed. Another option does exist and it is frequently employed by planners. The models can be run with data collected 10 or 15 years ago, to see how well they forecast present values.

Although planners do engage in such estimates, changing conditions and variations in the quality of data input to the models often result in accuracy levels that can differ substantially from one time period to another.

In addition, planning models used in transportation and air quality policy making, as we noted above, are actually a series of models used sequentially, such that the results of one model in the set become the inputs to the succeeding one. This implies that the accuracy of the entire set of models is what is relevant to policy making; but in practice, it is very difficult to estimate the accuracy of sequences of models in complex applications that extend over a time horizon of many years. An error in the output of one model may be magnified by a later model in the sequence, or it may not be, depending on the nature of the error and the mathematical characteristics of the particular models in the sequence.

It is reasonable to conclude that forecasting models applied to regional travel estimates and to air quality policy estimation cannot have their accuracy readily specified by a single overall indicator. The planners who use the models come to understand that they are more accurate for certain applications than for others, that they are better for comparing one broad policy option to another in terms of orders of magnitude, and that they can indeed be very poor analytical tools either for estimating the pollution reduction of a particular measure in a particular geographical sector or for estimating future traffic volumes on a particular roadway segment or transit route. After considering many of these factors, the committee of the Transportation Research Board, which was appointed to examine the relationships between expanding metropolitan highways and the implications for air quality and energy use, concluded quite unequivocally that,

> after examining the considerable literature on the relationships among transportation investment, travel demand, and land use as well as the current state of the art in modeling emissions, travel demand, and land use, the committee finds that the analytical methods in use are inadequate for addressing regulatory requirements. The accuracy implied by the interim conformity regulations issued by EPA, in particular, exceeds current modeling capabilities. The net differences in emission levels between the build and no-build scenarios are typically smaller than the error terms of the models. Modeled estimates are imprecise and limited in their account of changes in traffic flow characteristics, trip making, and land use attributable to transportation investments. The current regulatory requirements demand a level of analytic precision beyond the current state of the art in modeling.[7]

The court, which however still had to enforce the Clean Air Act, was less sympathetic than are experienced modelers who understand the complexity of the modeling exercise and know that models can produce idiosyncratic error patterns.

How the Court Ruling Changed
the Role of Planning Models

Although forecasting models have been widely used for decades in planning, for many of the reasons given above those who employ them have not generally taken their results to be literally true. Planners devise proposals for future transportation facilities and strategies for congestion management based on *several* kinds of information, of which modeling results are only one, and they have not traditionally had to defend any one of their technical methods or procedures in court. For example, when weighing courses of action, planners typically consider the performance of past transportation investments, insights from experiences of other transportation agencies, the responses of community groups, the preferences of funding agencies, and the views of governing boards as well as quantitative forecasts of policy outcomes. In addition, because of the long history of the evolution of planning methods, those who use them have interpreted the outcomes of transportation forecasting models as indicators of trends and of orders of magnitude: Does the model indicate that a subway is promising or not; does the model show that parking pricing is likely to be as effective as bridge tolls? Planners have rarely interpreted the models so literally in making policy recommendations that specific modal shares or particular numbers of trips on specific road segments or transit routes indicated by the model were taken to be the most likely future outcome of a public policy under scrutiny.

In the Bay Area litigation, however, the court treated the forecasts produced by planners' models rather differently. If a forecast showed that one transportation control measure reduced air pollutant emissions by 3.5 tons, that measure was literally taken to be half as useful as another control measure that the models indicated would reduce pollutant emissions by 7.0 tons. As indicated above, transportation planners, based on years of experience using these models for general policy guidance, might have recognized that the error terms associated with each forecast of reduced emissions might be as large or even larger than the estimates themselves. An estimate of 3.5

tons reduced by a transportation control measure in one model run could, for their purposes, be interpreted as sufficiently similar to another control measure's reduction of pollution by 7.0 tons to conclude that the two measures were not substantially different from one another in effectiveness. Planners used the models to determine whether one program would yield a 3-ton reduction in comparison with a 30-ton reduction in another plan or a 300-ton reduction in still another. Such a difference would be considered significant and worthy of careful attention.

Their understanding of the inaccuracy of the estimates produced by models caused planners to interpret such estimates less literally than the precise numerical values might suggest they could. But the court noticed that the modeling results were often stated with a fairly high level of precision and asked whether the numerical values produced by the planning models were the best available. If the answer was no, the court demanded that improved techniques be employed and the results be interpreted literally. On the other hand, if the answer was yes and these were indeed the best available estimates, the court felt justified in using the resulting values much more precisely in evaluating the effectiveness of alternative pollution control measures under the Clean Air Act than many planners felt were justified by the technical limits of the models. The court, less concerned with the limits of technical validity of the models, was more concerned with implementing the act's terms as best it could, using whatever sources of information it had available. This conflict between two views of the application of quantitative methods had to be resolved by the court in favor of the literal interpretation of the modeling results—the modeling studies were, after all, the only studies producing specific numerical estimates of future values of regulated pollutants.

Nowhere are the limitations of transportation planning models clearer than in the evaluation of transportation control measures. In Table 4.2, we listed the expected reductions in tons per day of the "2131" TCMs, as estimated by the MTC. Remember that the transportation control measures (TCMs) were contingency measures meant to remedy shortfalls in the necessary pollution reductions should the primary elements of the 1982 Plan fail to achieve the required reductions. The TCMs listed in the Plan survived from a substantially longer list, after elimination of measures for which no obvious sources of funds were available and elimination of measures that obviously lacked political support. The reductions shown in Table 4.2 were obtained by multiplying assumed percentage reductions (see Table 4.6) by the base emissions levels. This means, in short, that measures, which, at the margin, made the difference between attainment and nonattainment, were the result of sheer guesswork. There were few empirical estimates of any sort available on which to base estimates of percentage reductions from control measures such as employer audits (No. 23) or creating a new rail starts agreement (No. 16). Second, even in those areas where limited empirical data exist on which to base pollution reductions—raising bridge tolls (No. 13/14) or improving traffic signal timing (No. 24/25)—few if any such measures had ever been formally included in the regional transportation demand models. Despite this, the court used the quantitative estimates of their effects to arrive at a conclusion regarding the ability of the region to comply with the requirements of the Clean Air Act. Despite the fact that there were enormous differences in the likely ranges of accuracy of these estimates, the court gave them equal weight in determining their contributions to conformity.

The court's insistence on using the best available quantitative estimates of pollution reductions from the TCMs would seem to have two important and interrelated implications. First, it would appear to neutral observers that the MTC itself did not seriously contemplate the economic, political, and social costs and did not carefully weigh the air pollution benefits of the TCMs when they were initially included in the Bay Area Plan. The TCMs appear to be a catalog of measures, included to comply with a planning requirement, that were never subjected to detailed qualitative or quantitative analysis when the plan was drawn; in part, because at the time few thought they would ever become critical elements in a finding of compliance versus noncompliance. A very positive implication of the court action is that it gives notice to planning agencies all over the country that such plan elements are fulfilling statutory requirements and that agencies will be held to them. Consequently, this case implies that agencies should be careful not to include policy measures in plans without thorough analysis. In doing so, they risk court orders requiring governments to implement such measures, which in the end may not prove to be cost effective. In recent years, for example, employer trip reduction programs have come under scrutiny. Although they were listed as contingency measures in a number of regional air quality management plans, after evaluation in several metropolitan areas they are no longer regarded as very promising.[8]

The second implication of the court's decisions related to TCMs is equally important but more disturbing. In the future, improvements to travel demand forecasting and air pollution emissions and dispersion models could result in greater accuracy of, and improved confidence in, the air pollution consequences of transportation policy options. Improved capabilities of future models to address transportation control measures, however, will probably never yield the capacity to estimate all impacts of alternative policies with equal accuracy and confidence. Some policy options will always remain more risky than others; some will always involve actions that have fewer precedents than others and thus less empirical information on which to base estimates of impacts. The most creative policies are often the most speculative. Forcing planning agencies to rely to an ever greater extent on quantitative estimates of policy impacts might also in the end limit agencies' willingness to experiment with unproven strategies. On the one hand, it may be appropriate to take credit for contributions to attainment only when confidence can be placed in the ability of control measures to deliver their intended impacts. On the other, however, the biggest returns could in some cases result from as yet unproven options. Thus, although the court ruling seems to demand higher certainty and deeper study of TCMs before planning agencies incorporate them into their air quality management plans, it also seems to have the potential for stifling attempts at innovation.

Effects of Regional Growth on Transportation Models

In the case we discussed here, the court refused to require the MTC to develop a modeling approach that would account for potential changes in the overall level of regional growth on the assumption that building or not building transportation facilities would affect regional development. Instead, it allowed the agency to base its conformity assessments on the existing regional population and economic projections. It did, however, insist that the MTC consider the impact of congestion-relieving highway projects on regional development patterns and the potential for resulting changes in pollution levels. With respect to this critical question of whether additions to highway system capacity tend to alleviate or accentuate air pollution, the prestigious Transportation Research Board committee appointed to review this question could only equivocate. It stated in its final report,

> On the basis of current knowledge, it cannot be said that highway capacity projects are always effective measures for reducing emissions and energy use.

Neither can it be said that they necessarily increase emissions and energy use in all cases and under all conditions. Effects are highly dependent on specific circumstances, such as the type of capacity addition, location of the project in the region, extent and duration of preexisting congestion, prevailing atmospheric and topographic conditions, and development potential of the area.[9]

The EPA has now issued final conformity regulations covering the 1990 amendments that address this issue to some extent.[10] Although most states have submitted the required SIP revisions, which are in the process of being reviewed, in this transitional period there is as yet no clear track record on just how the EPA will interpret the amendments and its own regulations with regard to the new SIP submissions. Still, the regulations provide some insight into the agency's thinking on the matter of conformity.

As mentioned in Chapter 3, during the interim period, conformity is governed by the provisions in Section 176(c) and these EPA regulations (codified at 40 C.F.R. Part 51, subpart T).[11] Quantitative analyses are now explicitly required for all nonattainment and maintenance areas. Conformity determinations must be based on the most recent planning assumptions, including current estimates of population, employment, travel, and congestion along with background pollutant concentrations.[12] States must also use the most current motor vehicle emissions models in their assessments, though any new versions of EMFAC have to be approved by the EPA.[13] At a minimum, all areas must account for future growth in vehicle miles traveled (VMT) by considering historical growth trends in population.

During the interim period, both the regional transportation plan (RTP) and TIP for ozone and CO nonattainment areas must initially satisfy a "build"/ "no-build" test similar to the long-range conformity procedure that the MTC proposed to follow in this case. Although the agency suggested during the litigation that it should be required only to model emissions from the RTP through the year 1997, the new EPA regulations require metropolitan planning organizations (MPOs) to use a more extended analysis period. The analysis must at a minimum cover (a) the first milestone years (1995 for CO and 1996 for ozone), (b) the attainment year (if later than the milestone years, otherwise a year 5 years after the milestone years), and (c) the last year of the transportation plan's forecast period. The Intermodal Surface Transportation Efficiency Act (ISTEA) requires transportation plans to cover at least 20 years.[14] Interim analysis years may be no more than 10 years apart.[15] All plans and programs, and any project that does not come from a conforming plan or program, must contribute to emissions reductions.[16] This test requires highly sensitive estimates, comparing conditions with and without the proposed highway projects. In many cases, the differences will be quite small, often less than the error terms in the models. Once new SIPs are approved, the "build"/"no-build" test is no longer required but MPOs must still meet the emissions budget requirements.

The interim regulations define the "no-build" (or Baseline) scenario as the current transportation system (including projects under construction) projected forward to the future analysis years.[17] The "build" (or Action) scenario for the RTP is defined as the transportation system in each analysis year resulting from implementation of the proposed RTP, TIPs adopted under it, and all other expected "regionally significant" projects.[18] For the TIP analysis, it is the system that will result from implementing the proposed TIP and all other regionally significant projects within the time frame of the transportation plan.[19] For each analysis year, the MPO must estimate the CO and ozone emissions from travel on the "build" and "no-build" systems and compare the difference between the two.[20]

The criterion for interim conformity of transportation plans and programs is met if (a) emissions predicted in the "build" scenario are less than the emissions predicted from the "no-build" scenario in each analysis year, and if this can reasonably be expected to be true in the periods between the first milestone year and the analysis years; and (b) the "build" scenario contributes to some reduction in emissions from the 1990 emissions level.[21] These analyses will require projects to be more specifically defined to accurately estimate emissions.[22] MPOs that have already been using network models must continue to do so; otherwise, they can estimate regional emissions using methods that do not explicitly account for the influence of land use and transportation infrastructure on VMT and traffic speeds and congestion.[23]

A project that is *not* from a conforming RTP and TIP must also be shown to contribute to emissions reductions for CO and ozone.[24] Projects from a conforming RTP and TIP do not require any further regional ozone or CO analysis,[25] but all FHWA/FTA projects in CO nonattainment areas must either eliminate or reduce the severity and number of localized CO violations in the areas substantially affected by the project.[26] This hot spot analysis must also compare concentrations with and without the project, in combination with changes in background levels over time .

These requirements will certainly necessitate more sophisticated modeling techniques. The statute and regulations do not, however, specifically address the issue, raised above, regarding the effect of highway expansion on regional growth and future traffic and emissions levels. California's demographic forecasts anticipate that at least two thirds of all growth in the state will result from the natural increase of births over deaths, and only one third from an excess of in-migration over out-migration. Although highway construction will not likely affect the rate of natural increase, it was argued in the case that it is at least plausible, though unproven, that the rate of highway construction may affect migration trends. Assuming that one accepts the argument and wishes to take these effects into account, one possible approach would be to model different scenarios for each analysis year. For example, one scenario could involve the complete highway network and the forecast of all population and dwelling units for the target year. A second scenario could involve the "no-build" highway alternative and full regional population and housing growth. Scenario three might consist of the "no-build" highway alternative with reduced regional population and housing growth, reflecting the possibility that the region would be less attractive to future development under the "no-build" condition.

Adopting such an approach is not without its problems, though, because the modeling could lead to complex and confusing results. For instance, the model of full population growth and full highway network could lead to a finding of conformity with the Clean Air Act, the "no-build" transportation network with reduced population growth could also conform, whereas the "no-build" with full regional population growth might fail to conform. The result would pose a dilemma for public decision makers. In most states, growth policy is still a matter for local concern. MPOs and other planning agencies would probably not have the legal authority to require local governments to strictly adhere to measures that would restrict population growth in line with limited freeway construction. For its part, MTC believed that the public interest demanded that it be held to a transportation plan that would ensure conformity under the exogenous population forecasts, without being expected to implement unenforceable controls on population growth.

In fact, there are many historical instances in which rapid regional growth has taken place during periods of little of no highway construction; most major highway building programs have followed spurts of growth in population. In addition, plans that emphasize transit improvements over highway construction might be just as attractive to regional development. Although "no-build" or "reduced-build" highway construction alternatives would probably not lead to substantial reductions in regional growth, they could certainly lead to different growth patterns within a region, which could affect traffic congestion and travel patterns and in turn pollution levels.

The court in this case could find no credible evidence that transportation infrastructure investments have any discernible effect on the overall level of regional population growth or economic activity compared with other areas, and its decision was sound from a planning perspective. On the other hand, based on the available literature, the court did note that it was probable that the construction of new highway or transit facilities would affect the spatial distribution of economic activities and residences within a metropolitan area. Travel is determined to some extent by urban form, and urban form to some extent is determined by transportation investments; analytical methods are needed that address the simultaneity of these relationships. The implications of this assertion are that forecasts of future travel, pollutant emissions, and dispersion must be sensitive to alternative spatial arrangements of employment and residences and that these in turn must be sensitive to capital investments in transportation infrastructure. Conformity findings should, in principle, be based on forecasts of the spatial distribution of activities that are in equilibrium with forecasts of future travel.

Unfortunately, the state of the practice with respect to the mathematical modeling of urban form is probably even less advanced than the state of the practice of transportation modeling. As complex as models of travel and air quality need to be to realistically represent these policy variables and forecast them with reasonable accuracy, urban development models must be even more complex. They are based on economic and social principles and must recognize unique physical topography of urban areas, their patterns of in- and out-migration and of racial, ethnic, and class differences in employment and residential location, and much more. Many scholars who have reviewed the literature of urban development modeling and forecasting have concluded rather pessimistically that the power of these models falls far short of what policymakers appear to demand of them. Klosterman recently described the literature of urban development models as striving to move from an "art" to a "science."[27] An earlier and more exhaustive review of operational urban models in seven countries assessed their performance on a series of criteria all related in one way or another to their usefulness in policy making. The tests were conducted by measuring how well each model performed when applied to the specific city for which it was initially calibrated. The authors concluded rather guardedly that with respect to forecasts of urban form and the distribution of activities in space, the models had performed "plausibly," but in general they concluded that as support for predictions of future travel patterns, the models were less than satisfactory.[28] Although many authors agree that much has been learned in the past 20 years and that several operational models are now available, these models are difficult to adapt to new settings and offer highly variable results in practice.[29] In sum, urban development models do not offer forecasts of sufficient accuracy to require planning agencies to base regional air quality forecasts and determinations of conformity on them.

If a finding of conformity requires a planning study that looks 10 or 15 years ahead and considers future land use changes as well as future travel patterns, so that transportation investments and land uses interact as mutual causes and effects of one another, the requirement is placing a demand on the state of the practice that cannot be met by most planning agencies and consultants. Although new techniques of modeling, which might take advantage of tools such as Geographic Information Systems, may conceivably overcome these limitations in the future, a firm requirement that such models be used seems unrealistic in light of the performance record of currently available forecasting models.

The Consistency of Regional Travel Projections

The updated data, which appeared to show that pollution levels had risen significantly despite the adopted pollution control measures, put the court in the position of determining whether to hold the MTC liable for these apparent increases in emissions. The MTC's position, that it should not be accountable for the new estimates, was helped by the fact that the 1982 Plan was somewhat vague about whether the contingency provisions were intended to compensate for regional growth beyond that assumed in the initial baseline estimates. The defendants took advantage of that lack of clarity to hold their liability down to addressing shortfalls in their reduction strategy, not correcting any mistakes in estimating future growth.

With respect to those contingency provisions, the court ruled that the MTC was responsible for achieving specific overall residual levels of emissions, but that the reductions could be measured against the initial baseline estimates in the 1982 Plan rather than the more recent data. In other words, the court ruled that the responsible agencies had to develop additional pollution control measures if the initial measures chosen proved ineffective or, taken together, were inadequate to achieve the specific daily tonnage reductions called for in the Bay Area Plan. They would not, however, be held responsible if, despite their best efforts, those reductions did not in fact achieve the NAAQS. The proper solution in that case, consistent with the congressional mandate, would be to revise the original SIP. Defendants would be required to mitigate any anticipated increases in background emission levels following the initial attainment date, because those projections were incorporated into the original SIP, but they would not have to address any unanticipated increases due to growth. That would have to be dealt with in the new SIP required under the 1990 amendments.

As a practical matter, the court reached a reasonable result, especially given that a new plan must be prepared and adopted. On the other hand, the approach urged by MTC, which after all is responsible for demonstrating compliance with the Clean Air Act, raises some troubling questions, at least insofar as they insisted that the planning methodologies used in the revised emissions estimates did not correspond to the initial projections in the 1982 Plan. That agencies should keep up with the state of the art in monitoring and modeling practice can hardly be doubted. Still, the need to monitor past regulations to determine if they are meeting their declared objectives places some obligation on government officials to maintain consistency in practice. If the public is to have confidence in the government's ability to carry out its commitments under its planning programs, then it must produce data in a form suitable for that purpose.

The plaintiffs raised serious concerns about what could become a constantly shifting target, pointing out the difficulty facing public interest plaintiffs in retaining qualified experts to interpret agency data and generate sophisticated analyses needed to challenge agency decisions. The MTC contended that it could no longer even locate the computer programs and data used to produce the original Plan. Plaintiffs argued strenuously, but unsuccessfully, that the agency should be held to the newer data and that it represented only a refinement of earlier efforts not a completely different approach. The court had little choice but to accept the MTC's representations, though, because in fact the methodology used to determine pollution levels and the factors employed in that analysis differed significantly from the methodology employed in the original planning effort. Still, the issue of consistency in methodological approaches and comparability of data is one that needs to be addressed, particularly in light of the EPA's mandated use of the latest planning assumptions.

Setting aside for the moment the question of consistency in planning methods over time, there is still the issue of whether transportation planning agencies should be held responsible for the accuracy of their projections or, as the court ruling suggests, be entitled to rely on their initial trend line analysis even if it later proves to be flawed. The 1977 Clean Air Act did not directly address the question, but Congress realized that one reason for the widespread failure to meet federal air quality standards was the faulty model assumptions and inaccurate data used by many states. These increased uncertainty and understated the amount of emission reductions needed to reach attainment. As mentioned earlier, the 1990 amendments dealt with this question in part by requiring certain areas to monitor growth in VMT and vehicle trips (VTs)[30] and mandating that the SIPs for other more polluted areas affirmatively "offset any growth in emissions" from growth in VMT or VTs in such areas,[31] including adopting mandatory employer trip reduction programs.[32] In other words, these latter areas may not offset increases in vehicle-related pollution by improvements in stationary or other source controls.

Even nonattainment areas that are not subject to these strict requirements must account for any increase in pollution that may be due to growth in vehicle use. As discussed above, all Moderate and above ozone nonattainment areas will have to have submitted an attainment demonstration by November 15, 1993, showing a 15% reduction in ozone emissions by 1996.[33] Moderate areas must also attain the NAAQS by that date.[34] Areas listed as Serious or above for ozone have until at least 1999 to attain the NAAQS but must also demonstrate an average annual 3% reduction in emissions for every 3-year period thereafter until their applicable attainment dates, known as milestones.[35] All plans must show "specific annual reductions" in emissions needed to achieve the standards by the applicable dates.[36]

The target level of emissions is based on the 1990 base year inventory of mobile and stationary source emissions.[37] States must develop whatever control strategies are needed to meet that target. The EPA cautions air quality planning agencies that are used to thinking in terms of emissions reductions (that is, the difference between the 1996 projected inventory without controls applied and the 1996 target level) to be aware that the new regulation requires achievement of the actual target emission level. As the EPA explains,

The 1996 target level is dependent only on the 1990 emissions inventory, whereas the calculation of an emission reduction required relative to the current control strategy projection depends on the accuracy of the 1996 projection, which in turn depends on the estimate of future growth in activities. The assessment of whether an area has met the RFP requirement in 1996 will be based on whether the area is at or below the 1996 target level of emissions, and not whether the area has achieved a certain actual reduction relative to having maintained the current control strategy.[38]

In other words, MPOs will have to account for any growth in emissions and not, as the MTC argued in this case, just reduce baseline emissions. In addition to these ozone requirements, Moderate CO nonattainment areas will have to have achieved NAAQS by the end of 1995. Serious CO areas must attain the standards by the year 2000.[39] All areas with a design value above 12.7 parts per million (ppm) will have to have submitted a SIP revision by November 15, 1992, designed to ensure that the area achieves the "specific annual emissions reductions" necessary to demonstrate attainment by the applicable attainment date. Areas with a lower design value will not have to submit an attainment demonstration.[40] Contingency plans must be included for all areas to correct any shortfalls in meeting RFP targets or NAAQS.[41]

Beyond making the attainment and milestone demonstrations in the SIP, nonattainment areas must show that they actually achieved their "rate of progress" emissions reductions. Milestone compliance demonstrations are expected for Serious and above ozone areas by 1996 and every 3 years thereafter until attainment.[42] If any milestones are missed, contingency measures take effect automatically.[43] In addition, they must periodically determine whether their aggregate VMT, emissions, and congestion levels are consistent with the projections used in the SIP. If not, the state must submit a new SIP within 18 months, containing additional TCMs drawn from the Section 108(f) list.[44] Those CO nonattainment areas that must monitor VMT must also identify contingency measures to be undertaken if the number of annual miles traveled exceeds that predicted in the most recent forecast or if they fail to attain the NAAQS on time.[45] These areas must implement tracking plans to both monitor progress and trigger any required contingency measures.[46] Serious CO areas also will have to have established a milestone for December 31, 1995, identifying specific annual emissions reductions to be obtained by that date.[47]

These SIP requirements are supplemented with new conformity regulations that integrate attainment, RFP, and conformity. The EPA's new regulations require MPOs in all Moderate-and-above ozone areas and some

Moderate CO areas to establish a pollution emissions budget in the regional SIP and to periodically reassess whether their transportation plans and programs are consistent with it. Recall from Chapter 3 that the MTC argued forcefully that its conformity assessments should not be tied to the attainment levels in the SIP. The EPA's regulations do precisely that. They establish the conformity assessment procedure as the basic mechanism for ensuring that RFP is attained and maintained in accordance with the 1990 amendments. The EPA requires MPOs to use the latest planning assumptions, but the agency insists that given those assumptions, the emissions budget requirements must still be met.[48] These new mandates may make it harder for responsible agencies to sidestep their SIP commitments with arguments about apples and oranges.

The Emissions Budget Test

Submission of the control strategy SIP revision triggers the requirement for a new conformity determination for the RTP and TIP, based on the motor vehicle emissions budget required in the implementation plan.[49] The prior conformity status of the RTP and TIP remains valid for 12 months after submission of the revised SIP, after which it lapses and no additional projects may proceed unless a new conformity determination is made.[50] Until the SIP is approved, the plans and programs must also continue to meet the "build"/ "no-build" test to satisfy the Clean Air Act interim phase requirement that plans and programs contribute to annual emission reductions.[51] Areas with emissions budgets must perform a regional analysis to estimate emissions of hydrocarbons (HCs) and CO for the entire transportation system, including all regionally significant projects in the RTP and all other regionally significant highway and transit projects in the nonattainment or maintenance area planned within the time frame of the transportation plan.[52]

RTP and TIP Conformity

A regional emissions analysis must be performed for each Serious-and-above ozone nonattainment area and all Serious CO areas. Emissions from each pollutant in the attainment year, and in each analysis or horizon year after that, must not exceed the motor vehicle emissions budget for the attainment year.[53] The motor vehicle emissions budgets are the motor vehicle-related portions of the projected emission inventory used to demonstrate RFP for a particular year, as specified in the SIP.[54] Special rules apply to plans adopted after January 1, 1995, in these areas. They must be specific enough to be analyzed using state-of-the-art network-based travel demand models.[55] Planners must document demographic and employment factors influencing expected transportation demand, including land use forecasts. The emissions budget establishes a ceiling on emissions that cannot be exceeded by highway and transit vehicle emissions, irrespective of any difference between the area's current and forecasted population, employment, and travel demand and that forecasted in the SIP.[56] Unless the SIP expressly provides for a safety margin for conformity purposes, states cannot trade emissions between motor vehicle emissions and stationary sources to compensate for unexpected growth.[57]

No additional regional analysis is necessary for TIP conformity in areas with RTPs meeting the specific plan requirements for Serious and above CO and ozone nonattainment areas.[58] For all other areas, a regional emissions analysis must be performed that includes all projects in the proposed TIP, the RTP, and all other regionally significant highway and transit projects expected in the area in the time frame of the RTP.[59]

Project Conformity

Projects from a conforming RTP and TIP do not require any further regional ozone or CO analysis but, until the new SIP is approved, they must still meet the test for localized CO violations.[60] Emissions for projects that are not from a conforming RTP and TIP must, when considered with all other projects in the conforming RTP and TIP and all other regionally significant projects in the area, be consistent with the emissions budget in the SIP.[61]

The new rules represent a dramatic shift in pollution control strategies. The requirements relieve MPOs from monitoring the effectiveness of individual TCMs, and the determination of timely implementation is no longer based on a retrospective analysis of TCM effectiveness or whether each TCM had its predicted effectiveness unless the SIP specifically includes such a requirement. Under the new regulations, the transportation community is to be held responsible for implementing TCMs through the conformity process rather than merely as contingency measures. Given the difficulty of predicting TCM effectiveness or even measuring project-specific benefits once TCMs are implemented, MPOs are not responsible for achieving the emissions reductions predicted for each TCM. As the EPA explains, this is because any shortfall in emissions reductions is reflected in future conformity determinations through use of the latest planning assumptions, and because conformity is ultimately based on a comparison with an emissions budget. Shortfalls in emissions reduction from TCMs will either have to be offset by other measures in the transportation plan and TIP or else the transportation plan and TIP will not be in conformity and no highway projects can be approved.[62]

Although the regulations do not directly address the issue of growth, they do indirectly achieve a result similar to that advocated by the plaintiffs in this case. According to the EPA, due to the requirement that conformity determinations must use the most recent planning assumptions, it should be expected that conformity assessments will deviate from the SIP's assumptions regarding VMT growth, demographics, trip generation, and so on. The determination must nevertheless ensure that even given those revised assumptions, the emissions budget will be met.[63] The EPA's regulations governing conformity make clear that in approving plans and programs, local MPO's will be held accountable for any increased emissions:

> A regional analysis must estimate the emissions which would result from the transportation system if the RTP and TIP were implemented, and compare these emissions to the motor vehicle emissions budget identified in the SIP. If the emissions associated with the transportation plan and TIP are greater than the emissions budget, the RTP and TIP do not conform. *This may occur even though all TCMs in the SIP are properly implemented, as for example, if population and VMT growth are higher than predicted when the SIP was developed, motor vehicle emissions may exceed the SIP's budget for such emissions.*
>
> Under no circumstances may motor vehicle emissions predicted in a conformity determination exceed the pollutant-specific budget.[64]

Because conformity determinations must now be made whenever the RTP or TIP is amended[65] and at a minimum every 3 years,[66] there should be less opportunity for a region to drift out of compliance with the Clean Air Act. In addition, the transportation plan is subject to a conformity redetermination whenever the SIP is revised, as is the TIP if the RTP is amended.[67] If the conformity status is not redetermined as required, the RTP and TIP automatically lapse and individual projects cannot be approved. Requiring transportation forecasts to be accurate enough to establish an emissions budget in the SIP, and to make emissions comparisons between that budget and every RTP and TIP or risk losing highway projects, puts even more pressure on planners to improve the quality of their projections. Although the EPA does not explicitly require states to consider the effects of added roadway capacity on new development, the new regulations do deal at least indirectly with this issue. The requirement for continuous monitoring to confirm attainment and maintenance of the emissions budget, and the periodic conformity assessments, argues for modeling that is as accurate and realistic as possible, because some of the unexpected growth in emissions that must be accounted for could be due to feedback effects from capacity-increasing highway projects.

An Agenda for Transportation and Air Quality Analysis

Given that the court forced planners to confront the specificity of their modeling results, planners are now calling for a reconsideration of the role of quantitative models in planning and policy making. If planning models are to be interpreted literally and must be defended in court as though they were considered accurate to two or more decimal places, models that were previously seen as flawed but useful have lost some of their utility, and the pressure to develop a new generation of planning models is greater than ever.

In the late 1960s and early 1970s, transportation demand models were among the most advanced applications of computer modeling in the realm of public policy making. There was a lively research program dealing with the improvement of travel demand models, including specialists at several universities and several prestigious consulting firms. New ideas about travel demand analysis techniques were presented at most professional conferences on transportation, and the federal government promoted the development of planning models by funding research, disseminating advisory reports, distributing software packages, and sponsoring short courses in transportation analysis methods. At that time, theories of travel modeling and applications using a variety of new data collection techniques were proliferating, and one of the most important brakes on the state of the practice in planning agencies was limited computing capacity.

The situation has changed considerably during the past 20 years. Confronted with massive reductions in the scope of government during the 1980s, the federal role in research supporting new transportation planning methods, and in the concomitant training and other dissemination activities, has been reduced substantially. Although a small cadre of dedicated academics continues to promote new approaches to travel demand modeling and analysis, it lacks the continuous base of funding support that was characteristic of the field decades ago. In addition, although federal regulations still suggest that the planning methods be used by metropolitan area transportation planning agencies, the level of expertise and the sophistication of planning modeling software available varies greatly from region to region. In many parts of the country, there has been little progress or improvement in agency capabilities for more than a decade. Several consulting firms have perfected their own versions of transportation planning software and market applications of their software internationally. They have an economic incentive to adapt their software to new applications and specifications, but their profits are maximized by making minimal changes in the basic core of their modeling software packages. Thus, experts in transportation planning widely agree not only that the state of the practice has advanced very slowly during the past decade or so but also that some metropolitan areas are using planning models so primitive as to constitute an embarrassment to those areas. Critically, the rate of progress has fallen far behind the demands being placed on transportation planners by new situations, including those related to air quality, congestion management planning, increased attention to goods movement, and other issues that characterize transportation planning in the 1990s. Although at one time progress in the state of the practice was limited by computing capacity, recent advances in computing capacity make a wide range of alternative approaches to modeling possible, but they are being pursued in very few metropolitan areas.[68]

An important outcome of the Bay Area lawsuit has been a renewed focus on the need to improve the technical capabilities of travel demand forecasters, especially strengthening the linkages between transportation and air quality analysis. Although some government officials, many academics, and some planning practitioners working in the most advanced agencies have called for a new generation of planning methods for some time, the lawsuit galvanized interest and support for initiatives in planning methods to an extent that had not been seen in more than 15 years. Piecing together funding contributions from the Federal Highway Administration, the Federal Transit

Administration, the U.S. Environmental Protection Agency, and the U.S. Department of Energy, a new Travel Model Improvement Program (TMIP) was launched in 1993.[69] The program is envisioned by those shaping it to stretch over more than 5 years and to require a budget of more than $25 million, although far less has so far been actually committed to the program.[70] Early in the program, several transportation modeling researchers were commissioned simultaneously to develop research agendas and proposals, and the results were widely disseminated and discussed at transportation conferences.[71] At a research conference held in Fort Worth, Texas, in August of 1994, the TMIP was introduced to the national community of transportation scholars and practitioners, with more than 130 people in attendance. The discussions were lively and at times major disagreements surfaced over the future direction of the TMIP, but all participants shared a sense of urgency about the importance of the task of updating the scope and range of transportation planning methods.

The TMIP is structured around four major activities, each of which is referred to as a *track*. Although some of the tracks are as yet funded at only very low levels, the intent of the federal sponsors and the review panel advising them is to give each of the areas serious attention over the coming several years. The ability of the federal sponsors to actually carry out this plan is challenged by recent proposals by the Clinton Administration and the new Republican congressional majority to further reduce federal spending, so it is difficult to predict at this moment whether the program will succeed. The first of the four tracks intends to emphasize *outreach*. Work conducted under this track will help practitioners improve their existing planning procedures to be consistent with currently desirable practice. Outreach activities will consist of a program of training, technical assistance, research coordination, and the creation of a clearinghouse for research findings. The second track will focus on *near-term improvements*. This track will consist of a program of technical activities to help metropolitan planning organizations and state departments of transportation elevate current practice to state of the art. These efforts will implement planning model improvements already developed but not widely included in current transportation, land use, and air quality activities at metropolitan areas and at the state level. This track intends to improve the state of the practice mostly among agencies that seem to be lagging behind those considered most advanced. The third track is aimed at *longer term improvements*. It has already undertaken major research and development initiatives intended to create fundamentally new approaches to travel and land use forecasting. Envisioned to take at least 5 years and expected to absorb a large part of the entire program's funding, this track has the goal of producing a new generation of travel and air quality planning models that will overcome many of the limitations discussed earlier. Thus far, a controversial decision has been made to fund the development of TRANSIMS, a software package proposed by the Los Alamos National Laboratory that includes the direct simulation of household travel patterns and their air quality consequences.[72] The work to date by the TRANSIMS team is promising but is opposed by many transportation researchers who favor the continuation of other trends in transportation research. The administrators of the program are trying to foster greater understanding on the part of the Los Alamos researchers of recent trends in transportation research and are attempting to foster greater collaboration through subcontracts to established transportation and air quality modelers, who might then work directly with the TRANSIMS team. The final track is devoted to improvement in *data collection* in support of all the other tracks. Efforts in this track will identify, design, and develop improved data collection procedures that will meet decision makers' current and future needs. The results of the Bay Area lawsuit are among the many determinants of the need for improved data collection capabilities.

Conclusion

Transportation and air quality planners have known for more than a decade that their techniques were falling systematically behind the demands of such federal mandates as those incorporated in ISTEA and the Clean Air Act Amendments. Academic spokespeople, thoughtful agency representatives, and consultants were pointing out the new demands being placed on aging software by the changing policy environment in which transportation planning and air quality conformity were being addressed in an increasing number of metropolitan areas. The case reviewed in this book became a landmark that has been widely cited by advocates of improved approaches because it added a sense of urgency and some political clout to their calls for improvements. There is now great fear among planning agencies that their current methods fall short of their assigned tasks, and that similar lawsuits in other cities and states might soon embarrass them and require them to perform tasks their planning tools are incapable of doing well. It is impossible as of this date to be certain that the TMIP will either receive the funding it seeks or succeed within the time frame of 5 years. Whether the TMIP succeeds or fails, a major adjustment will have to be made in our way of doing regional transportation and air quality planning. The state of the practice simply falls short of the demands of federal requirements, and we must adjust our methods and expectations in order to develop consistency between agencies' technical capabilities and the requirements of federal laws and regulations.

Notes

1. 60 Fed. Reg. 27,028 (May 23, 1995). Authority for the redesignation to attainment status is provided in Section 107(d) of the Clean Air Act. 42 U.S.C.A. § 7407(d) (West 1995). To attain the redesignation, the state must submit a SIP revision that provides for maintenance of the NAAQS for at least 10 years. A subsequent SIP revision is due 8 years later, covering the remaining 10 years. 42 U.S.C.A. § 7505a (West 1995). EPA approval of the redesignation begins the "maintenance period." See 40 C.F.R. § 51.392 (1994).

2. The redesignation to attainment status relieves the state from having to submit a "15% SIP" for ozone by November 15, 1993, though it must still maintain the federal standard over the next 20 years.

3. Bay Area Air Quality Management District, *Making History: A Forty Year Success Story, 1955-1995* (undated).

4. D. E. Boyce, N. D. Day, and C. MacDonald, *Metropolitan Plan Making: An Analysis of Experience with the Preparation and Evaluation of Alternative Land Use and Transportation Plans,* Monograph No. Four (Philadelphia: Regional Science Research Institute, 1970).

5. Transportation Research Board, *Expanding Metropolitan Highways: Implications for Air Quality and Energy Use,* Special Report 245 (Washington, DC: Transportation Research Board, 1995), 29-31.

6. Boyce, Day, and MacDonald, op. cit., chapter 4.

7. Transportation Research Board, op. cit., pp. 5-6.

8. G. Giuliano, G. K. Hwang, and M. Wachs, "Employee Trip Reduction in Southern California: First Year Results," *Transportation Research A,* 27A (1993): 125-137; Institute of Transportation Engineers, Technical Council Committee, *Employee Trip Reduction Programs: An Evaluation* (undated); Apogee Research, Inc., *Costs and Effectiveness of Transportation Control Measures (TCMs): A Review and Analysis of the Literature,* Prepared for the Clean Air Project of the National Association of Regional Councils, 1994.

9. Transportation Research Board, op.cit., p. 7.

10. Criteria and Procedures for Determining Conformity to State or Federal Implementation Plans of Transportation Plans, Programs, and Projects Funded or Approved under Title 23 U.S.C. or the Federal Transit Act, 58 Fed. Reg. 62,188 (November 24, 1993) [hereinafter Conformity Rule] (codified at 40 C.F.R. Part 51 (1994)). The rule applies only to plans, programs, and projects developed, funded, or approved under Title 23 of the Federal Transit Act. The criteria for determining conformity for all other federal actions are contained in 40 C.F.R. Part 51, Subpart W. See 58 Fed Reg. 63,214 (November 30, 1993).

11. The interim period was to run only until the conformity SIP revision required by Section 176(c)(4)(C) was approved. 42 U.S.C.A. § 7506(c)(3) (West 1995). See Chapter 3, note 42 supra. The EPA administratively extended the time for submitting the conformity SIP revision until November 24, 1994. Conformity Rule, supra note 10, at 62,189. The EPA has also extended the interim period requirements until the control strategy SIPs are submitted, or the deadline for submission of the control strategy SIP revision, whichever is earlier, at which time the transitional period begins and the emissions budget test applies. Id. at 62,191 (see 40 C.F.R. § 51.392 (1994)). The SIP conformity assessment provisions take effect once the control strategy revision is approved by the EPA.

12. 40 C.F.R. § 51.412 (1994).

13. 40 C.F.R. § 51.412(a)&(b) (1994).

14. 23 U.S.C.A. § 134(g) (West Supp. 1995).

15. 40 C.F.R. § 51.436(b) (1994).

16. 40 C.F.R. §§ 51.436(a), 51.438(a), 51.440 (1994).

17. 40 C.F.R. § 51.436(c) (1994).

18. 40 C.F.R. § 51.436(d) (1994). A "regionally significant project" is a (nonexempt) transportation project on a facility that serves regional transportation needs (such as access to and from the area outside the region, major activity centers in the region, major planned developments, or transportation terminals) and would normally be included in the modeling of a metropolitan area's transportation network. At a minimum, all arterial highways and all fixed guideway transit facilities that offer an alternative to regional highway travel must be included. Id. § 51.392.

19. 40 C.F.R. § 51.438(d) (1994).

20. 40 C.F.R. §§ 51.436(e); 51.438(e) (1994). Emissions in milestone years that are between the analysis years may be determined by interpolation.

21. 40 C.F.R. §§ 51.436(f); 51.438(f) (1994).

22. 40 C.F.R. § 51.452(a) (1994). The regulations require RTPs in all areas to identify the future transportation system in sufficient detail to perform the required conformity determinations. The emissions analysis must include all FHWA/FTA projects proposed in the transportation plan and TIP and all other regionally significant projects. Projects that are not regionally significant are not required to be explicitly modeled, but VMT from such projects must be estimated in accordance with reasonable professional practice.

23. 40 C.F.R. § 51.452(c) (1994). These areas may extrapolate existing or future VMT by considering growth in population and historical growth trends for VMT per person, adjusting for future economic activity, transit alternatives, and other TCMs.

24. 40 C.F.R. § 51.440 (1994). The criteria can be satisfied by a regional emissions analysis that includes the project in the "build" scenario. If the project is a modification of a project currently in the RTP or TIP, the "no-build" scenario must include the project with its original design and scope and the "build" scenario must include the project with its new design concept and scope.

25. A project is considered to be from a conforming transportation plan if (a) it is one of the projects required to be included in the plan and its design concept and scope have not changed significantly from when it was included in the plan, or (b) it is identified in the plan and is consistent with the policies and purpose of the plan and will not interfere with other projects specifically included in the plan. A project is considered to be from a conforming program if (a) the design concept and scope of the project were adequate at the time of the TIP conformity determination to determine its contribution to the TIP's regional emissions and it has not changed significantly and (b) written commitments to implement any project-level emissions mitigation

or control measures described in the TIP are obtained from the project sponsor. 40 C.F.R. § 51.422 (1994).

26. 40 C.F.R. § 51.434(a) (1994). The criterion is satisfied if existing localized CO violations will be eliminated or reduced in severity and number as a result of the project. Quantitative modeling based on EPA guidelines (40 C.F.R. part 51, appendix W (Guideline on Air Quality Models (Revised)(1988), supplement A (1987) and supplement B (1993), EPA publication no. 450/2-78-027R)) is required for all projects involving or affecting (a) sites identified in the SIP as sites of current violation or possible current violation; (b) intersections that are Level-of-Service (LOS) D, E, or F or that will change to LOS D, E, or F because of increased traffic volumes related to the project; (c) the three top intersections in the area based on highest traffic volume; or (d) the three worst intersections for LOS. Id. § 51.454(a)(1)-(4). In all other areas, the criterion may be satisfied by a qualitative analysis if consideration of local factors clearly demonstrates that existing CO violations will be eliminated or reduced. Otherwise, a quantitative analysis, which must reflect reasonable and common professional practice, will be required. Id. § 51.434(c); see also § 51.454(b). The CO hot spot analysis must include the entire project and may be performed only after the major design features that will significantly impact CO concentrations have been identified. Id. § 51.454(c). It must also be consistent with the assumptions in the regional emissions analysis. Id. § 51.454(e).

27. R. E. Klosterman, "An Introduction to the Literature on Large-Scale Urban Models," *Journal of the American Planning Association* 60 (Winter 1994): 41-44.

28. F. V. Webster, P. H. Bly, and N. J. Paulley, Eds., *Urban Land Use and Transport Interactions: Policies and Models,* Report of the International Study Group on Land Use/Transport Interaction (Aldershot, UK: Transport and Road Research Laboratory, 1988).

29. M. Wegener, "Operational Urban Models: State of the Art," *Journal of the American Planning Association* 60 (Winter 1994): 17-29.

30. 42 U.S.C.A. § 7512a(2)(A) (CO areas with a design value above 12.7 ppm) (West 1995). See General Preamble for the Implementation of Title I, 57 Fed. Reg. 13,498, 13,520, ¶ III.A.4.(n) (April 16, 1992) [hereinafter General Preamble]. The plan revision must provide for annual updates of the forecasts to be submitted to the EPA along with reports regarding the extent to which the forecasts proved to be accurate.

31. 42 U.S.C.A. § 7511a(d)(1)(A) (Severe ozone); § 7512a(b)(2) (Serious CO) (West 1995). The EPA interprets these provisions to require that volatile organic compounds (VOC) emissions never be higher in one year than during the year before,

> When growth in VMT and vehicle trips would otherwise cause a motor vehicle emissions upturn, this upturn must be prevented. The emissions level at the point of upturn becomes a ceiling on motor vehicle emissions. This requirement applies to projected emissions in the years between the submission of the SIP revision and the attainment deadline and is above and beyond the separate requirements for the RFP and the attainment demonstrations.
>
> General Preamble, supra note 30, at 13,521-2 (April 16, 1992).

32. By November 15, 1992, these areas must require each employer of 100 or more persons to increase average passenger occupancy per vehicle in peak period work trips by not less than 25% above the average vehicle occupancy for all such trips in the area. Employers must submit compliance plans within 2 years and convincingly demonstrate compliance within 4 years. 42 U.S.C.A. § 7511a(d)(1)(B) (West 1995).

33. 42 U.S.C.A. § 7511a(b)(1)(A) (West 1995).

34. 42 U.S.C.A. § 7511(a)(1) (West 1995).

35. 42 U.S.C.A. § 7511a(c)(2)(B) (West 1995). The SIP revision and 3% annual RFP demonstration are due by November 15, 1994. See Notice of Proposed Rulemaking, 58 Fed. Reg. 3769 (January 11, 1993).

36. 42 U.S.C.A. § 7511a(b)(1)(A) (West 1995).

37. General Preamble, supra note 30, at 13,508. The 1990 motor vehicle baseline emissions is obtained by multiplying the 1990 VMT by a hypothetical emissions factor for 1996. This adjusts for emissions that would be eliminated through fleet turnover. This figure is added to nonmotor-vehicle HC emissions, and the total is multiplied by 0.85 to obtain the 1996 target level of emissions.

38. Id.

39. 42 U.S.C.A. §7512(a)(1) (West 1995).

40. 42 U.S.C.A. § 7512a(a)(7) (West 1995).

41. 42 U.S.C.A. §§ 7502(c)(9) (West 1995).

42. 42 U.S.C.A. § 7511a(g)(1) (West 1995). SIPs in these Moderate and above ozone areas must project annual progress that will result from their control strategies (including appropriate implementation schedules and expected emissions reductions) to meet the 15% requirement and the 3% milestones. The primary means of demonstrating this rate of progress requirement will be through periodic inventories submitted every 3 years. General Preamble, supra note 30, at 13,512, ¶ III.A.3.(d); 13,518, ¶ III.A.4.(g); see also § 7502(c)(3) (requirement for periodic inventories). Failure to meet the milestone may also result in reclassification to a higher classification, implementation of specific measures adequate to meet the next milestone, or the adoption of an economic incentive program. 42 U.S.C.A. § 7511a(g)(3) (West 1995).

43. 42 U.S.C.A. § 7511a(c)(9) (West 1995). The EPA believes that contingency measures providing additional emissions reductions of up to 3% of the emissions in the adjusted base year inventory in the year following must be in the 1993 SIP. General Preamble, supra note 30, at 13,511, ¶ III.A.3(c).

44. 42 U.S.C.A. § 7511a(c)(5) (West 1995). ("In considering such measures, the State should ensure adequate access to downtown, other commercial and residential areas and should avoid measures that increase or relocate emissions and congestion rather than reduce them.")

45. 42 U.S.C.A. § 7512a(a)(3) (West 1995). The EPA indicates that an appropriate contingency measure for exceeding a VMT forecast would be sufficient VMT reductions or emissions reductions to counteract the effect of one year's growth in VMT while the state revises its SIP to provide for attainment. These measures may offset either the excess VMT or the additional CO emissions in the areas that are attributable to the additional VMT. General Preamble, supra note 30, at 13,532, ¶ III.B.2.(b).

46. CO nonattainment areas must submit a revised inventory every 3 years, beginning September 30, 1995. 42 U.S.C.A. § 7512a(a)(5). For CO areas, the EPA regulations call for a tracking plan to annually update estimates of VMT in future years and to verify that contingency measures are being implemented if either the actual VMT estimates for the previous year or any new VMT forecasts for any year until the attainment year exceed any earlier forecasts in the SIP. The updates must discuss the extent to which such forecasts proved accurate. General Preamble, supra note 30, at 13,533, ¶ III.B.2.(a)&(f).

47. 42 U.S.C.A. § 7512a(d)(1) (West 1995). If a state fails to submit an adequate demonstration within 9 months, the state must submit a plan to implement an economic incentive and transportation control program sufficient to achieve the specific annual reductions in CO emissions by the attainment date in the SIP. Id. § 7512a(d)(3).

48. The EPA rejected suggestions that it require an assessment of both the degree to which key assumptions deviate from those used in the SIP and an evaluation of the impact of any

deviation on the area's ability to reach its SIP target, reasoning that the targets had to be met in any event. Conformity Rule, supra note 10, at 62,210.

49. 40 C.F.R.§§ 51.428(a); 51.430(a) (1994).

50. 40 C.F.R. § 51.448(a) (1994). For control strategy, SIP revisions submitted after November 24, 1993, RTP and TIP conformity must be determined according to the transitional period rules within 12 months from the deadline for submission. During this 12-month period, the existing RTP and TIP are still valid and projects from them may proceed, provided that the NEPA process is completed and the project is found to conform.

51. 40 C.F.R. § 51.436 (1994); see Conformity Rule, supra note 10, at 62,191; 42 U.S.C.A. § 7506(c)(3)(A)(iii) (West 1995).

52. 40 C.F.R. § 51.428(b) (1994).

53. 40 C.F.R. § 51.428(c)(2) & (c)(3) (1994). If emissions budgets are established for years after the attainment year, emissions in each analysis year or horizon year must be less than or equal to the emissions budget for that year, if any, or for the most recent budget year prior to the analysis year or horizon year.

54. Unless otherwise indicated, the SIP's estimate of future highway and transit emissions used in the milestone or attainment demonstration is the motor vehicle emissions budget. Conformity Rule, supra note 10, at 62,195; see also General Preamble, supra note 30, at 13,558.

55. 40 C.F.R. § 51.452(b)(1). The RTP must specifically describe the transportation system for several horizon years, none of which may be more than 10 years apart, including the attainment year if it is in the time span of the transportation plan, and the last year of the plan's forecast period. 40 C.F.R. § 51.404(a)(1). For each horizon year, the plan must quantify the demographic and employment factors influencing expected transportation demand, including land use forecasts. Id. § 51.404(a)(2)(i). The modeled highway and transit system in these specific plans must contain all regionally significant additions or modifications to the existing transportation network that will be operational in the horizon years. Each added or modified highway segment must be sufficiently identified in terms of its design concept and design scope to allow modeling of travel times under various traffic volumes, consistent with the modeling methods for areawide transportation analysis in use by the MPO. Transit facilities, equipment, and services must be defined to allow modeling of transit ridership. The descriptions of additions and modifications to the transportation network must be specific enough to show that there is a reasonable relationship between expected land use and the envisioned transportation system. Id. § 51.404(a)(2)(ii).

56. The EPA explains this requirement as follows,

> The EPA interprets these provisions to mean that the combination of highway capacity expansion, highway extensions, support for transit, and TCMs in the transportation plan and program must result in vehicle emissions that are not in excess of those contained in the SIP's demonstration of RFP and attainment, despite any difference that may exist between the areas's current and forecasted population, employment and travel demand and those that were assumed at the time of SIP preparation and adoption. In other words, the conformity provisions envision that the SIP will create an emissions budget (for the criteria pollutant and its precursors) for highway vehicles, and that the transportation planning process will be required to produce plans and programs that will result in emissions within that budget.
>
> General Preamble, supra note 30, at 13,557.

57. 40 C.F.R. § 51.456 (1994).

58. 40 C.F.R. § 51.430(b) (1994). The TIP must be consistent with the conforming transportation plan and must demonstrate that (a) the TIP contains all projects that must be started in

the TIP's time frame to achieve the highway and transit system envisioned in each of the transportation plan's horizon years, (b) all TIP projects that are regionally significant or are part of the specific highway or transit system in the RTP's horizon year, and (c) the design concept and scope of each regionally significant project in the TIP is not significantly different from that described in the RTP.

    59. 40 C.F.R. § 51.430(c) (1994).

    60. 40 C.F.R. § 51.434(a) (1994). See note 26 supra. In any event, no project may cause or contribute to any new localized CO violations or increase the frequency or severity of any existing CO violations in any CO nonattainment or maintenance area. 40 C.F.R. § 51.424(a) (1994). For purposes of SIP development, quantitative modeling is required for all intersections that are LOS D, E, or F or that will change to LOS D, E, or F because of increased traffic volumes related to the project; the three top intersections in the area based on highest traffic volume; and the three worst intersections for LOS. EPA Guidelines for Modeling Carbon Monoxide from Roadway Intersections, November 1992.

    61. 40 C.F.R. § 51.432(a) (1994).

    62. Conformity Rule, supra note 10, at 62,198.

    63. Id. at 62,210.

    64. Id. at 62,194-5 (emphasis added).

    65. 40 C.F.R. § 51.400(b)(2) & (c)(2) (1994).

    66. 40 C.F.R. § 51.400 (b)(4) & (c)(4) (1994). If RTP/TIP conformity determinations are not made every 3 years or within the grace period following a trigger, the RTP/TIP conformity status lapses and no new project-level determinations may be made.

    67. The conformity of existing transportation plans must be redetermined within 18 months of (a) November 24, 1993, or (b) EPA approval of a SIP revision (or promulgation of a federal implementation plan) that establishes or revises a transportation-related emissions budget or adds, changes, or deletes TCMs. 40 C.F.R. § 51.400(b)(3) (1994). A TIP conformity redetermination is also required within 6 months of the adoption of a new RTP or an RTP revision (unless it merely adds or deletes exempt projects). Id. subsection (c)(3).

    68. *Travel Demand Forecasting Processes Used by Ten Large Metropolitan Planning Organizations,* Institute of Transportation Engineers Committee 6Y-53, August 18, 1993.

    69. U.S. Department of Transportation, *Travel Model Improvement Program* (undated)

    70. U.S. Department of Transportation, *Funding the Travel Model Improvement Program(TMIP)/TRANSIMS* Draft (December 1, 1993), 4-5.

    71. B. D. Spear, *New Approaches to Transportation Forecasting Models: A Synthesis of Four Research Proposals* Final Report. Prepared for the U.S. Department of Transportation, Federal Highway Administration, 1993.

    72. TRANSIMS uses microsimulation to forecast traffic and provides detailed information for estimating motor vehicle emissions. It is expected to provide the technical capabilities needed to produce information for transportation policy analysis and air quality impact assessment.

APPENDIX A*
Conformity Assessment

Section 176(c) of the Clean Air Act includes a provision that:

> No metropolitan planning organization designated under section 134 of title 23, United States Code, shall give its approval to any project, program, or plan which does not conform to a plan approved or promulgated under Section 110.

MTC is the MPO for the Bay Area. The above requirement will be satisfied in the following manner:

- The Regional Transportation Plan will be assessed yearly at the time it is being amended (September) to determine if it complies with the Air Quality Plan. This assessment will include: (1) an evaluation of the continued support of the TCMs, and (2) a determination of the air quality impacts if the RTP amendments. The Commission will make a formal determination of conformity after this assessment.
- The Transportation Improvement Program (TIP) will be reviewed each year to determine its compliance with the Air Quality Plan. All transportation projects with Federal funding must be in the TIP. The review will include: (1) an assessment of the implementation of the adopted TCMs in the TIP, and (2) an assessment of major highway projects to determine if they will adversely affect emissions. This review will be included as a chapter of the TIP.
- Individual project application currently undergo an environmental assessment by the MTC. This will also serve to ensure conformity.

*Appendix A and Appendix B are excerpted from the *December 1982 Bay Area Air Quality Plan,* Appendix H, prepared by the Association of Bay Area Governments, Bay Area Air Quality Management District, and Metropolitan Transportation Commission.

APPENDIX B
Contingency Plan

EPA guidelines require a contingency plan for transportation measures which will be implemented if reasonable further progress is not achieved. This contingency plan contains three elements:

a. *List of transportation projects with potentially adverse air quality impacts which will be delayed while the air quality plan is being revised.*

The MTC believes that providing a list of specific projects in this plan is impractical for two reasons:

- the project list would change depending on the year that the RFP target is not met
- the environmental documentation on projects in the later years of the Transportation Improvement Program (TIP) is not yet available in most cases.

Accordingly, if the RFP target is not met, MTC may delay certain categories of projects in the TIP if they are shown to have significant adverse impact on air quality. The criteria for delaying projects and specific projects to be delayed will be determined following the initial public hearing under section b. The categories which may be delayed include:

- Freeway Congestion Relief Projects (HB 42)*
- Freeway Traffic Service Projects (HB 43)
- Conventional Highway and Expressway Operation Improvement Projects
- New Connection (HB 44) and Cross-Traffic Improvements (HE 11)
- Upgraded Facilities (HE 12)
- Lane Additions (HE 13)
- New Highways (HE 14)
- Projects funded by the Federal Aid Urban Program which increase roadway capacity.

b. *Process for determining/implementing additional TCMs*

In July of each year, an RFP report will be submitted to EPA. Part of this report will be a review of the status of implementation of the adopted TCMs. A second portion would assess the growth in vehicle travel in the region. If a determination is made that RFP is not being met for the transportation sector, the MTC will adopt additional TCMs within 6 months of the determination. These TCMs will be designed to bring the region back within the RFP line.

MTC will conduct the following process within the 6 month time frame:

- hold an initial public hearing to solicit comments on projects to be delayed and suggestions on additional controls;
- review progress made in implementing controls adopted in the Air Quality Plan;
- analyze additional controls;
- hold a final public hearing prior to adoption of additional measure.

c. *Annual Inspection/Maintenance Program*

The MTC will support legislative authorization for an annual I/M Program.

---

*Designation refers to Caltrans category.

230

About the Authors

Mark Garrett is a Los Angeles attorney specializing in land use planning and environmental issues. He holds a BA degree in urban studies from Case Western Reserve University and both a JD degree and an MA degree in urban planning from UCLA. He has coauthored an article on vested development rights for the *UCLA Journal of Environmental Law and Planning* and a forthcoming article, to be published by Fannie Mae, based on a recently completed study of rent control in four California cities. His latest article, a review of current literature on the history of American planning, appears in the student journal *Critical Planning*. As a practicing lawyer he has advised local governments on zoning, subdivision and redevelopment law, inverse condemnation, and growth management issues. He has also served as a consultant to the City of Oxnard, overseeing the preparation of the city's 2020 General Plan. He is currently pursuing a PhD, with his major research focus on property rights and land use regulation, particularly the impact of growth controls on affordable housing.

Martin Wachs is Professor of Urban Planning and Director of the Institute of Transportation Studies at UCLA, where he has been a member of the faculty since 1971 and also served three terms as head of the urban planning program. He holds a bachelor's degree in civil engineering from the City University of New York, and MS and PhD degrees in transportation planning from the Civil Engineering Department at Northwestern University. Before coming to UCLA, he was an assistant professor at Northwestern University and the University of Illinois at Chicago. In 1986, he received a Distinguished Planning Educator award from the California Planning Foundation and a Distinguished Teaching Award from the UCLA Alumni Association. He is the author or editor of four books and has written more than 90 published articles on transportation planning and policy, including the transportation needs of elderly and handicapped people, fare and subsidy policies in urban transportation, the problem of crime in public transit systems, and methods for the evaluation of alternative transportation projects. He has also done historical studies of the relationship between transportation investments and urban form in the early part of the 20th century and on ethics in planning and forecasting. Recently, his writings have dealt with the relationship between transportation, air quality, and land use. He has also been studying the effects of parking pricing and congestion pricing on traffic and travel decisions. The recipient of a Guggenheim Foundation fellowship, he has twice won research fellowships from the Rockefeller Foundation and has been awarded the Pike Johnson Award, given annually to the best research paper presented at the annual meeting of the Transportation Research Board.